Mathematical Conundrums

Want to sharpen your mathematical wits? If so, then *Mathematical Conundrums* is for you. *Daily Telegraph* enigmatologist, Barry R. Clarke, presents over 120 fiendish problems that will test both your ingenuity and persistence. Between these covers are puzzles in geometry, arithmetic, and algebra (there is even a section for computer programmers). And, for the smartest readers who wish to stretch their mind to its limits, a selection of engaging logic and visual lateral puzzles is included. Although no puzzle requires a greater knowledge of mathematics than the high school curriculum, this collection will take you to the edge. But are you equal to the challenge?

Features

- High school level of mathematics is the only prerequisite.
- Variety of algebraic, route-drawing, and geometrical conundrums.
- Hints section for the lateral puzzles.
- Warm-up exercises to sharpen the wits.
- Full solutions to every problem.

Barry R. Clarke has published over 1,500 puzzles in *The Daily Telegraph* and has contributed enigmas to *New Scientist*, *The Sunday Times*, *Reader's Digest*, *The Sunday Telegraph*, and *Prospect* magazine. His book *Challenging Logic Puzzles Mensa* has sold over 100,000 copies. As well as a PhD in Shakespeare Studies, Barry has a master's degree and academic publications in quantum physics. He is now working on a revised theory of the hydrogen atom. Other skills include mathematics tutor, filmmaker, comedy-sketch writer, cartoonist, computer programmer, and blues guitarist! For more information please visit http://barryispuzzled.com.

AK Peters/CRC Recreational Mathematics Series

Series Editors
Robert Fathauer
Snezana Lawrence
Jun Mitani
Colm Mulcahy
Peter Winkler
Carolyn Yackel

For more information about this series please visit: https://www.routledge.com/AK-PetersCRC-Recreational-Mathematics-Series/book-series/RECMATH?pd=published,forthcoming&pg=2&pp=12&so=pub&view=list

Mathematical
Conundrums

Barry R. Clarke

CRC Press
Taylor & Francis Group
Boca Raton New York London

CRC Press is an imprint of the
Taylor & Francis Group, an **informa** business

Designed cover image: Barry R. Clarke

First edition published 2023
by CRC Press
6000 Broken Sound Parkway NW, Suite 300, Boca Raton, FL 33487-2742

and by CRC Press
4 Park Square, Milton Park, Abingdon, Oxon, OX14 4RN

CRC Press is an imprint of Taylor & Francis Group, LLC

ISBN: 978-1-032-41478-2 (hbk)
ISBN: 978-1-032-41458-4 (pbk)
ISBN: 978-1-003-35827-5 (ebk)

DOI: 10.1201/9781003358275

Typeset in Optima
by SPi Technologies India Pvt Ltd (Straive)

Contents

Contents

Preface

Welcome to *Mathematical Conundrums*! This collection presents over 120 of my best enigmas that have appeared in national newspapers and magazines. They rely primarily on intelligence and persistence, and need no greater knowledge of mathematics than that taught in high school. However, their difficulty lies in the level of ingenuity and persistence required. So who might derive pleasure from this book? It should be ideal for the intelligent high school student who is seeking a challenge beyond the established curriculum, or the university science student who wants to keep their mind in shape, or in general, anyone with a working knowledge of arithmetic and algebra who enjoys thinking. In addition, many of these problems will be excellent preparation for the UKMT Mathematical Challenge and the MAA American Mathematics Competitions for schools. Full solutions are given at the end of each chapter. Good luck!

Barry R. Clarke
Oxford, UK

Introduction

The first book I recall seeing on recreational mathematics was Martin Gardner's *Mathematical Puzzles and Diversions* [1]. I was 14 years old at the time and found it in the school library as a hardback copy, with a picture of an arrangement of pentominoes on its green dust cover. The chapter that fascinated me the most was on Sam Loyd (1841–1911) who Gardner billed as "America's Greatest Puzzlist." Since I was interested in conjuring tricks, Loyd appeared to me at the time as an almost supernatural puzzle magician. One of his earliest puzzles, which he claimed to have invented at the age of 9, ran as follows [2].

It is told that three neighbours, who shared a small park, as shown in the sketch (Figure 1.1) had a falling out. The owner of the large house (A), complaining that his neighbour's chickens annoyed him, built an enclosed pathway from his door to the gate at the bottom of the picture (A). Then the man on the right (B) built a path to the gate on the left (B), and the man on the left (C) built a path to the gate on the right (C), so that none of the paths cross, and each man has an exit opposite his door.

Can you draw the paths?

The solution is given at the end of the chapter. In his article, Gardner states that "Loyd's most interesting creation is the famous '14–15' or 'Boss' puzzle." This consists of a four-by-four array of squares in which there are 15 numbered blocks 1–15 with one empty square. The puzzle is set up with the blocks in serial order except for the 14 and 15 which are juxtaposed, the

DOI: 10.1201/9781003358275-1

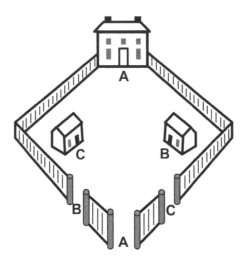

Figure 1.1   Sam Loyd's pathway puzzle.

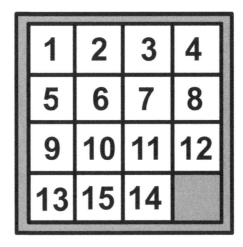

Figure 1.2   Initial configuration of the 14–15 or Boss puzzle.

empty square being left in the bottom right corner (see Figure 1.2). The vacant space is to be used to slide all the blocks into serial order, again leaving the space in the right-hand bottom corner of the grid.

In January 1896, Loyd offered a prize of $1000 in *The Illustrated American* claiming the puzzle as his own invention. He knew he would never have to pay out because a mathematical paper had been published six years earlier demonstrating its impossibility [3]. Not only that, when *The New York Times* reported on the puzzle at the height of its craze on 22 March and 11 June

1880, Sam Loyd received no credit even though he lived in New York. It transpired that the puzzle wasn't his. Noyes Palmer Chapman, a postmaster from Canastota in New York, had filed a patent application dated 21 February 1880 on his "Block Solitaire Puzzle," a 4×4 grid with 15 sliding blocks that could be removed. The application was rejected, likely on the grounds that a patent for a 6×6 grid with 35 sliding pieces had earlier been granted on 20 August 1878 to Ernest U. Kinsey. Apart from the number of sliding blocks, the main difference seemed only to be that Kinsey's "Puzzle Blocks" were held in place by tongues and grooves while Chapman's blocks could be tipped out [4]. Sam Loyd also "borrowed" the ideas for his "P. T. Barnum's Trick Donkeys" [5] and "Get Off the Earth" puzzles from elsewhere. I only learned these facts many years later with great disappointment, and the discovery left me with the conclusion that Sam Loyd excelled more at shameless self-promotion than inspired invention. As he himself said, "people don't care for my puzzles unless they can have them with my name on them" [6].

A contemporary of Sam Loyd also had misgivings about his claims to priority. Henry E. Dudeney (1857–1930) actually accused Loyd of stealing his puzzles and presenting them under his own name. Dudeney was a master of dissection problems and his most celebrated discovery was the rearrangement of the parts of an equilateral triangle into a square (see Figure 1.3). This he called the "Haberdasher's Puzzle" which he published in the *Weekly Dispatch* on 14 June 1903. Dudeney was a prolific compiler, and his main puzzles column appeared in *The Strand Magazine* and ran for 20 years from March 1908.

Although Martin Gardner excelled more at popularisation than compilation, he was the first to present the problem of dissecting a square into eight acute triangles. This he announced in *Scientific American* in February 1960 [7, 8]. As a university physics student, I managed to succeed in producing nine acute triangles using a pentagon in one corner but was disappointed to learn that my solution was not original.

Figure 1.3   Dudeney's equilateral triangle to square dissection.

The inspiration for my own puzzle compilation began with winning the *New Scientist* Enigma prize for 27 March 1986. This was a puzzle entitled "In the bag" set by Christopher Maslanka, the compiler for *The Guardian*, and involved discovering the numbers of red, white, and blue eggs in a bag under various selection conditions. At the time, I was a mathematics PhD student at University College Swansea, a degree I neglected to complete due to my greater interest in quantum physics. I recall that the puzzle, which required the method of exhaustion, took me four hours to crack with just pen and paper. A year later, I bought Chris Maslanka's *The Pyrgic Puzzler* [9] and delighted in the humorous names and drawings of his many quirky characters. For example, here is puzzle No. 9.

Professor Pembish knows that (with the plug in) his bath fills in three minutes if the cold tap only is turned on full, and in four minutes if the hot tap only is turned on full. With the plug out (and the taps off) the bath empties in two minutes. One day he finds to his horror that Mrs Oldham has mislaid the plug, but he is intelligent enough to realise that he can still fill the bath with both taps turned on full.

How long will it take to fill?

Around 2002, I came across the lateral thinking puzzles of Lloyd King who has invented many ingenious puzzles that demand an alternative inter-pretation of a visual scene to arrive at the solution [10, 11]. I have met only a handful of people who excel at this type of thinking, and it definitely requires an inventive mind. Something resembling this type of skill is employed for the solution of a geometrical problem, when the imaginative addition of a line leads to new relationships.

Lloyd King's "Snowman" puzzle is shown in Figure 1.4. The problem is as follows.

Which month is indicated in this wintery scene: April, May, June, July, or August?

Of course, it is important to give a reason for your answer! In Chapter 8, I give examples of my own puzzles of this type. Some have criticised this puzzle form as "mind-reading," especially if no hints are given, but my recommendation is that if you see no way forward, don't feel defeated

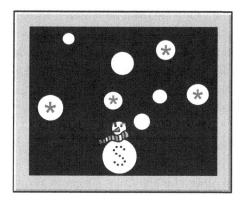

Figure 1.4   The "Snowman" puzzle.

or cheated. Simply look up the solution and try to admire its construction as puzzle art.

My first published puzzle was "One for the road" which appeared in *New Scientist* on 14 July 1988 [12]. It involved five drunks who were trying to push their broken-down car home. They were so inebriated that some pushed at the front of the car, and others opposed them by pushing at the back. The puzzle was to work out the whole-number strength of each man from the given conditions. My second puzzle also appeared in *New Scientist*, this time as "Sum secret" on 26 August 1989. It was an example of the set of digital deletion sums that I had worked out (see Figure 1.5).

In the sum shown, the first row added to the second gives the third, the fourth subtracted from the third gives the fifth, and the fifth added to the

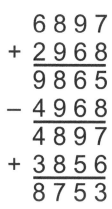

Figure 1.5   Digital deletion puzzle.

sixth gives the seventh. One digit can be erased from each row (not necessarily the same position in each row) and the gaps can be closed up to leave three columns of digits, then a second digit can be rubbed out in the same way to give two columns, then a third to leave one column, so that a valid sum remains each time. The three sets of seven digits erased (read down the columns) respectively reveal three numbers.

What are the three numbers?

In 1989, I was employed by *The Daily Telegraph* as one of the four members of the Brain Twister team under the editorship of Val Gilbert. Rex Gooch, Angela Newing, David Singmaster, and I provided a weekly puzzle in rotation. Then in the mid-1990s, I became the sole compiler with the assistance of Jacqui Harper's excellent cartoon illustrations. To date the *Telegraph* has published over 1,500 of my enigmas. One of my earliest inventions was the "Mix-and-match" logic puzzle which first appeared in the *Telegraph* on 21 May 1994. An example appears below with four rows and three columns of items. Each item is in the correct column, but only one item in each column is in the correct row. The given facts about the correct order lead to the solution.

(1) Freeman was one place below the boxer from Boston.
(2) The fighter from Seattle was one place above the boxer nicknamed Iron.
(3) McCool was two places below the contestant nicknamed Sugar.

*Can you find the correct nickname, surname, and hometown for each
.   position?*

My most difficult puzzle of this type has seven rows and four columns of items which has approximately 6×10^{14} possible arrangements [13]. An example appears as puzzle 7.17 in Chapter 7.

Table 1.1 Four by three mix-and-match puzzle

	Nickname	Surname	Hometown
1	Rocky	Tryson	Boston
2	Sugar	Holyhead	Seattle
3	Basher	McCool	Texas
4	Iron	Freeman	New York

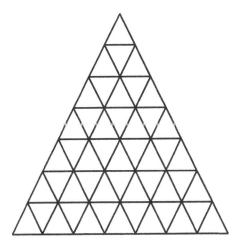

Figure 1.6   A master triangle of side length n = 7.

I have also taken an interest in recreational mathematics problems. One investigation that particularly interested me appeared in my book *Mathematical Puzzles and Curiosities* [14]. It involves the triangle shown in Figure 1.6. For this I managed to deduce formulae that predict both the number of triangles and the number of quadrilaterals of any size in a master triangle of side length n. For the total number of triangles T_n we have the following.

$$T_n = \frac{n(2n+1)(n+2)-\delta_n}{8}, \qquad \delta_n = \begin{cases} 0 \text{ for } n \text{ even} \\ 1 \text{ for } n \text{ odd} \end{cases} \tag{1.1}$$

The total number of quadrilaterals Q_n is given by

$$Q_n = \frac{n(n^2-1)(n+2)(2+\varepsilon_n)}{8}, \qquad \varepsilon_n = \begin{cases} \dfrac{n(n-2)}{2(n^2-1)} \text{ for } n \text{ even} \\[2mm] \dfrac{n^2-3}{2n(n+2)} \text{ for } n \text{ odd} \end{cases} \tag{1.2}$$

For example, for $n = 3$, we find $\delta_3 = 1$ and $\varepsilon_3 = 1/5$, so $T_3 = \dfrac{3(7)(5)-1}{8} = 13$ and $Q_3 = \dfrac{3(8)(5)(11/5)}{8} = 33$.

These are some of the puzzles that have captured my interest during my enigmatic journey. For the collection of my conundrums that follow, I hope you get as much pleasure solving them as I had compiling them.

Solutions

Sam Loyd's neighbour puzzle

Figure 1.7 appeared in the January 1908 issue of *The Strand Magazine* [15].

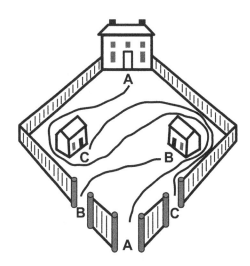

Figure 1.7   Solution to Sam Loyd's pathway puzzle.

Square dissection into eight acute triangles

Points P and Q lie outside the grey semi-circles which shows that the angles are acute (see Figure 1.8).

Professor Pembish's bath

The bath fills in 12 minutes. Let the volume of the bath be V, and the rates for the cold tap, hot tap, and drainage be r_c, r_h, and r_d, respectively. Then $r_c = V/3$, $r_h = V/4$, and $r_d = V/2$. Let the time for the bath to fill with all three rates working together be t. Then $(r_c + r_h - r_d)t = V$. Substituting the rates leads to cancellation of the V and $(1/3 + 1/4 - 1/2)t = 1$. So $t = 12$.

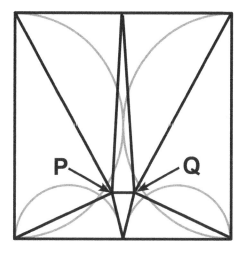

Figure 1.8   Square dissection into eight acute triangles.

Snowman

The month is June. The scene represents our solar system. The body of the snowman is the Sun and the head is Mercury. The planets with asterisks are the Earth, Jupiter, Uranus, and Neptune. We then take their initial letters. The added clever touch about this puzzle is that the title pun "snowman" can be interpreted as "it's no man," that is, what you are looking at is not a man.

Digital deletion sum

The numbers are 7868753, 8656865, and 9999988.

Mix-and-match

Table 1.2 Solution to the four by three mix-and-match puzzle

	Nickname	Surname	Hometown
1	Sugar	Holyhead	Boston
2	Basher	Freeman	New York
3	Rocky	McCool	Seattle
4	Iron	Tryson	Texas

References

[1] Gardner, Martin. *Mathematical Puzzles and Diversions from Scientific American*. G. Bell & Sons, 1961.

[2] Bain, George Grantham. 'The Prince of puzzle-makers'. *The Strand Magazine*. Vol. xxxiv (December 1907): 772.

[3] Johnson, William Woolsey, and Story, William E. 'Notes on the "15" puzzle'. *American Journal of Mathematics* 2 (December 1879): 397–404.

[4] Slocum, Jerry, and Sonneveld, Dic. *The 15 Puzzle*. The Slocum Puzzle Foundation, Beverley Hills, California, 2006, pp.100–103.

[5] Gardner, Martin. *Mathematical Puzzles and Diversions from Scientific American*. Pelican Books, 1979, p.86.

[6] Bain, George Grantham. 'The Prince of puzzle-makers'. *Strand Magazine*. Vol. xxxiv (December 1907): 773.

[7] Gardner, Martin. 'Mathematical games'. *Scientific American*. Vol. 202 (February 1960): 150.

[8] Gardner, Martin. *New Mathematical Diversions*. The Mathematical Association of America, 1995, p.40.

[9] Maslanka, Chris. *The Pyrgic Puzzler*. Kingswood, 1987.

[10] King, Lloyd. *Test Your Creative Thinking*. A *Times* book. Kogan Page, 2003.

[11] King, Lloyd. *Amazing "Aha" Puzzles*. Lulu.com, 2004.

[12] Clarke, Barry R. *Puzzles for Pleasure*. Cambridge University Press, 1993, p.75.

[13] Clarke, Barry R. *Extreme Logic Puzzles*. Puzzlewright Press, 2015, p.62.

[14] Clarke, Barry R. *Mathematical Puzzles and Curiosities*. Dover Publications, 2013.

[15] 'Solutions of puzzles and problems in the Christmas number'. *The Strand Magazine*. Vol. xxxv (January 1908): 114.

Mind sharpeners

2.1 Banging a drum

The Little Drummer Boy plays 120 beats a minute when he is happy, but only 60 when he is sad. In the previous minute, he was sad for twice as long as he was happy.

How many beats did he play in that minute?

2.2 Money for nothing

Each Christmas, as his benevolence declines with age, Scrooge gives away £4 less than he did the previous year. Thirty years ago he gave away a square even number between £100 and £200, no two digits being equal.

How much does he give away this year?

2.3 Switched on

While fitting wall insulation at Muckrake mansions, Phil McCafferty has found three hidden switches inside the walls. He discovers that they control three disused light bulbs in the house in three one-to-one connections, that is, each switch controls only one bulb, and each bulb is controlled by

DOI: 10.1201/9781003358275-2

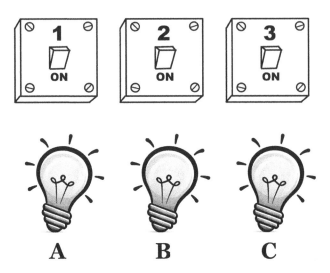

Figure 2.1 Light switches and bulbs.

only one switch. On reporting his discovery to the owner Sir Giles Crust, he makes one statement about each switch.

Switch 1: "Controls bulb B."
Switch 2: "Controls bulb A or C."
Switch 3: "Controls bulb A or B."

However, Phil cannot prevent himself lying at random and it turns out that only one statement is true.

Can you match the switches to the bulbs?

2.4 Humble pie

Tiny Tum eats 8 mince pies and 4 apple pies. An apple pie weighs three times the weight of a mince pie. Altogether, he ate a total weight of 280 g.

What is the weight of one apple pie?

2.5 Number feast

Good King Wenceslas had been invited to his friend Stephen's banquet. This was to be two days before the day that was after the day that was six days before the day when Christmas day was yesterday.

What date was the feast of Stephen?

2.6 Feeling sheepish

Since Carol the shepherd found it impossible to watch her flock by night, three-fifths of them ran away. If four-sevenths had gone instead, she would have lost seven fewer sheep.

How many sheep were originally in the flock?

2.7 The drinking dwarfs

Over Christmas, Sleepy and Dopey drink a total of 35 pints of beer. For every two pints Sleepy drinks, Dopey drinks five.

How many pints does Dopey drink?

2.8 Dicey Business

Sad Sally has just finished with her boyfriend, so she looks up the telephone numbers of six ex-boyfriends numbered 1 to 6 in her address book. Unsure which one to contact she throws a dice and lets chance decide. The number that comes up might be described as follows.

(1) not a prime number
(2) less than 4
(3) even number
(4) square number
(5) odd number

However, exactly two statements are false.

What number is thrown on the dice?

2.9 Hey ho!

Two of the seven dwarves are drunk, so the remaining five dwarves set off to work.

How many ways can they form a straight line so that Bashful is next to Doc and Sneezy is next to Happy?

2.10 Tournament Town

One of the 12 roads in Tournament Town has been closed due to a snowdrift (see Figure 2.2). A route from the white house to the dark grey house need not use every road.

How many such routes are possible without passing along any road (shown light grey) more than once?

Figure 2.2   Roads in Tournament Town.

2.11 Can you dig it?

A Trogg can dig twice as fast as a Grimble. A duo-spade can remove twice as much soil as a spade. It takes 2 Troggs and 4 Grimbles twelve hours to remove 64 m³ of soil from the ground each using a spade.

How long does it take 4 Troggs and 2 Grimbles to remove 40m³ each using a duo-spade?

2.12 The digital age

A veteran of several Twitter wars, Grandpa was now being interviewed by *Demob Matters Magazine* to canvas his views on various social media issues. However, he was never one to fire directly at the target. When asked his age he gave the following enigmatic response.

(1) It consists of two different digits 1–9.
(2) The difference between the digits is not a prime number.
(3) My age is divisible by 3.
(4) The sum of the digits is greater than 6.
(5) At least one digit is a triangular number.
(6) Reversing the digits of my age gives a number less than my age.

How old is Grandpa?

2.13 Acting the goat

At Four-Footed Farm there are a certain number of goats in a field. This turns out to be a square number less than 500. The field is surrounded by a fence and entrance to it is through a gate. It is goat-cleaning day and none are looking forward to their bath. So the following sequence of events occurs.

(1) Three escape by crawling under the fence.
(2) One third of those left jump out over the fence.
(3) Then two change their mind and crawl back in under the fence.
(4) One fifth of the remainder stampede out of the gate when the farmer opens it.

How many goats are left in the field?

2.14 Letting off steam

The drivers of two steam engines, the Eggspress (light grey) and the Funderbolt (dark grey), each 5 m long, are in dispute about who has the faster machine (see Figure 2.3). So they decide to race each other on parallel straight tracks between two signals A and B set 100 m apart. The engines take a long run up towards signal A and the race starts just as the back end of each engine passes A simultaneously. At this point they travel at different constant speeds E and F, respectively, which they maintain throughout the race. It's a thrilling encounter, and the front end of the Funderbolt is one train length away from the rear of the Eggspress when the front end of the Eggspress reaches signal the Eggspress.

What is the ratio of speeds E: F?

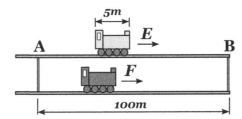

Figure 2.3   The Eggspess and the Funderbolt.

2.15 Cubic sagacity

Professor Neuron was staring at the cube of water in his fish tank (see Figure 2.4). The inside of the tank was cuboid shaped, with the base measuring $a \times a$ square feet, and the water having an internal height of a feet. The professor placed a square number b of identical cuboid bricks in the tank, each having an area of one square foot in contact with the inside of the tank base. The water subsequently rose to an internal height of c feet without exceeding either the tank or the brick height.

The numbers a, b, and c are digits from 1 to 9 inclusive with no two digits being identical.

What are the values of a, b, c?

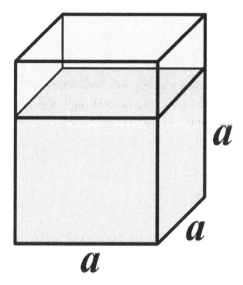

Figure 2.4   Water in the fish tank.

2.16 Test in times

In preparation for building a house wall, Juan the bricklayer decided to practise his skills on a test wall. He informed his colleagues that he could build it alone in 25 hours. However, Rafael announced that he could lay bricks at two-thirds of the rate that Juan can lay them. So instead of Juan working alone, they decided to work together on it continuously.

How many hours did it take them to build the test wall together?

2.17 Horse play

General Dobbin, the military leader of the equine fraternity, was adamant that the average number of horse flies encamped on any horse at any moment was at least a two-digit number. Squadron Leader Clump, the leader of the horse-fly world, disagreed and stated that it was only a single digit 1–9. So Clump called a news conference in the Pose Garden at the Flight House and made six statements about the digit.

(1) Less than 5.
(2) Prime number.
(3) Divisible by 3.

(4) Odd number.
(5) Does not immediately follow a prime.
(6) Square number.

Unfortunately, Clump was incapable of consistently telling the truth and so not all the statements are true. The correct digit is exactly divisible by the number of false statements.

What is the digit?

2.18 Al jabber

Young Al had just finished reading his book *Difficult Sums* as his grandfather Grumble entered the room holding a watering can.

Said Al excitedly, "Seven years ago, I was a third of the age you were, when you were twice the age I will be in seven years time." Grumble was used to his grandson's irrelevant utterances so he ignored him. He was far more interested in watering his pot plant.

Neither has reached the age of 100 and the sum of their ages is a square number.

How many years ago was Grumble twice the age Al will be in seven years' time?

2.19 Antwit's age

Twenty years ago, Antwit said, "one year before I'm thrice the age I am now, I'll be the product of Babble's age now and the age Crumble will be three years after he is twice the age he is now." Each of their ages 20 years ago was a single digit 2–9 inclusive, no two digits being equal.

How old is Antwit at this present moment?

2.20 Prickly problem

When Holly was twice the age Ivy was 15 years ago, Ivy was half the age Holly will be in 12 years' time.

How many years ago was that?

Solutions

2.1 Banging a drum

The drummer boy plays 80 beats. The boy is sad for 2/3 minutes and happy for 1/3 minute. So he plays 60(2/3) = 40 beats when he is sad and 120(1/3) = 40 beats when he is happy.

2.2 Money for nothing

Scrooge gives away £76. Let the amount this year be x. The only possible square even number is 196 and thirty years ago he gave away 4(30) = £120 more. So, 30 years ago he gave away $x + 120 = 196$. So $x = 76$. Alternatively, we could see it as an arithmetic series with first term 196 and difference −4. We require the 31st term if we are to advance 30 years. Since the nth term is −4n + 200 then with $n = 31$ we have 76.

2.3 Switched on

Switch 1 controls bulb C, switch 2 controls bulb B, and switch 3 controls bulb A. Only the last statement is true. If the first statement is true then the other two are false. This allows switch 1 to connect to B and switch 2 to B which is invalid. If the second statement is true then switch 2 controls A or C. Also the first and third are false so that switch 1 controls A or C, and switch 3 controls bulb C. Bulb B cannot be lit which is invalid. Finally, if the third statement is true then switch 3 controls A or B. The first two statements are false, so switch 1 controls A or C, and switch 2 controls B. So switch 3 controls A and switch 1 controls C.

2.4 Humble pie

Tiny Tum's apple pie weighed 42 g. Let the weight of an apple pie be A and a mince pie be M. Then $8M + 4A = 280$ and $A = 3M$. Substitution of the second into the first gives $M = 14$ and $A = 42$.

2.5 Number feast

The feast was on 19th December. Working backwards from Christmas Day 25th, we need 25 + 1 − 6 + 1 − 2 = 19.

2.6 Feeling sheepish

There were 245 sheep in the flock. Let the number be x. Then $3x/5 = 4x/7 + 7$. Multiplying by 35 gives the solution.

2.7 The drinking dwarfs

Dopey drinks 25 pints. Let Dopey's number of pints be D and Sleepy's be S. Then $S + D = 35$ and $S = 2D/5$. Substituting the latter into the former and multiplying by 5 gives $7D = 175$ and the answer follows.

2.8 Dicey business

The number thrown is 4 and statements (2) and (5) are false. There must be three true statements. If (1) is true the number is 1, 4, or 6; for (2) 1, 2, or 3; for (3) 2, 4, or 6; for (4) 1 or 4; and for (5) 1, 3, or 5. We need the number that appears three times and this is 4.

2.9 Hey ho!

There are 24 ways. Link the pairs together so that instead of arranging 5 items there are 3 items to arrange: two connected pairs and a single dwarf. There are 6 ways of arranging these 3 items. Inside each connected pair there are two ways round they can be. This gives a total of $6(2)(2) = 24$.

2.10 Tournament Town

There are 8 routes. When two arrows occur then their paths must be followed consecutively (see Figure 2.5).

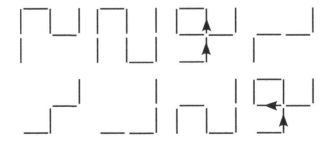

Figure 2.5   Possible routes for Tournament Town.

2.11 Can you dig it?

It takes 3 hours. Let t be the time, v the volume of soil, n the number of men (where a Trogg does the work of two men and a Grimble counts as one), and s the spade size (where a spade has value 1 and a duo-spade 2). Then using a constant of proportionality k, we can write $t = kv/(ns)$. In the first scenario, $t = 12$, $v = 64$, $s = 1$, and $n = 8$ (2 Troggs is worth 4), so $k = 3/2$. In the second scenario, $v = 40$, $s = 2$, and $n = 10$ (4 Troggs do the work of 8 men). So $t = (3/2)(40)/(20) = 3$.

2.12 The digital age

Grandpa is 93. From condition (3), their digital sum is divisible by 3. Condition (2) leaves the pair of digits as (1, 5), (2, 4), (3, 9) or (4, 8), with the order undetermined. From condition (4), we are left with (3, 9) or (4, 8). After applying (5) we have (3, 9). Finally, condition (6) gives the age.

2.13 Acting the goat

There are 120 left. Let the square number be x. Then $(4/5)[(2/3)(x - 3) + 2] = L$, an integer. Only the square number $x = 225$ allows a division by 5 in the final calculation.

2.14 Letting off steam

The ratio of speeds is E:F = 19:17. In the same time, the back end of the Eggspress travels 95 m (remember its back end signals the starting time while its front end marks the finishing time) while the back of the Funderbolt travels 85 m. Note that the Funderbolt's front end is one train length away from the back of the Eggspress, so its back end (which is what we are using to measure distance travelled) is two train lengths away. So, 95:85 = 19:17.

2.15 Cubic sagacity

The values are $a = 6$, $b = 9$, and $c = 8$.
   The water has volume a^3 and when the b bricks are inserted, it has the same volume but expressed as $(a^2 - b)c$. Equating these leads to $c = a^3/(a^2 - b)$. We now choose the possible square numbers for b, namely 1, 4, 9, and

run through the possible values of a for each, to check if c is from 1 to 9 inclusive. There is no suitable value of c for the values $b = 1, 4$. However, for $b = 9$ we arrive at $c = 216/27 = 8$ when $a = 6$. In fact, this is the only solution.

2.16 Test in times

It takes them 15 hours. Let there be X bricks in the wall. Let J be the number of bricks Juan lays per hour. Then $25J = X$. Let the unknown time for them both to complete the wall be T. Then $T(J + 2J/3) = X$. So $T = X/(5J/3) = 75/5 = 15$ hours.

2.17 Horse play

The correct digit is 4. For each condition we list the possible digits. For example, condition (1) has 1, 2, 3, 4. Then for each digit we count the number of statements it appears in. If a digit is the correct answer, this count would be the number of true statements. We subtract each count from six to find the number of false statements for each digit. Only the digit 4, which if true would generate 4 false statements, is exactly divisible by the number of false statements.

2.18 Al jabber

It was two years ago. Let Al's and Grumble's ages now be A and G, respectively. Then $A - 7 = (2/3)(A + 7)$ which gives $A = 35$. Let the number of years ago that Grumble was $2(A + 7)$ be x. So $G - x = 2(A + 7) = 84$, that is $G > 84$, and $A + G > 119$. Considering their sum being a square, if $A + G = 121$, $G = 86$, and if $A + G = 144$, $G = 109$. Since neither age exceeds 100, Grumble is 86 and so $x = 2$.

2.19 Antwit's age

Antwit is 29 at present. Twenty years ago, Antwit was 9, Babble was 2 and Crumble was 5. Let Antwit, Babble, and Crumble's ages 20 years ago be A, B and C, respectively. Then we have $3A - 1 = B(2C + 3)$. Since no digit can be 1, then B cannot be 1 and so $3A - 1$ is not prime. This leaves $A = 3, 5, 7$, or 9. For $A = 3$, $3A - 1$ is 8 but $2C + 3$ cannot be 2 or 4. For $A = 5$, $2C + 3$ must be 7 (it cannot be 2), but then $B = C = 2$ (invalid duplication). For $A = 7$, we want the factors of 20 so $(B, 2C + 3) = (2, 10)$, $(4, 5)$, or $(5, 4)$ but no $C > 1$ can be found for any of them. This only leaves $A = 9$ where $3A - 1 = 26$. Then $(B, 2C + 3) = (2, 13)$. So $A = 9$, $B = 2$, and $C = 5$.

2.20 Prickly problem

It was six years ago. Let Holly and Ivy's ages be H and I, respectively, and let the number of years in the past be x. Then $H - x = 2(I - 15)$ and $I - x = (H + 12)/2$. Making x the subject in each gives $x = H - 2I + 30$ and $x = I - H/2 - 6$. Multiplying the second of these by 2 and adding the two equations leaves $3x = 18$.

3 Geometry

3.1 Calculating feat

George the giant wants to plant a circular lawn but has no measuring equip-
ment. So, he decides to walk 14 paces in a straight line from A to B, and
then a whole number of paces in a straight line from B to C (see Figure 3.1).
His route A–B passes through the circle centre, and B–C is a tangent to the
circle. Also, points A and C lie on the circle circumference, George's pace
is consistently 7 metres, and the lawn radius is a prime number of metres.

What is the diameter of the lawn in metres?

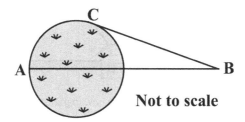

Figure 3.1   The giant's circular lawn.

DOI: 10.1201/9781003358275-3

3.2 Paper trial

In the village of Little Apens, near Borehamwood in UK, Drudge the news-paper delivery boy rises at 6 every morning to deliver the morning papers. The round always begins at the newsagent (white house) and finishes when he arrives back home (black house). His delivery requirement can vary from day to day, but the number of customers (grey houses) who require a deliv-ery is from one to four inclusive. So on some days, certain houses might receive no visit.

During a round, although Drudge can repeat a visit to a customer's house, he makes it his rule never to traverse the same road (shown grey) between two adjacent houses more than once. Of course, on some days certain roads might not be traversed at all.

How many different routes are possible for Drudge's delivery round?

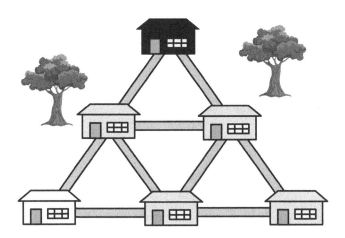

Figure 3.2   Possible paths for the newspaper delivery route.

3.3 The bee and the box

Billy the bee is trapped in a closed cardboard box with the dimensions shown above. Rather than fly blind, he walks in a straight line from corner A to corner B. Eventually, someone cuts opens the box along the edges and he escapes. A few days later, Billy returns to find that the box faces have been unfolded and laid out flat into a net. This time Billy lands at point A and walks in a straight line from A to C.

What is the difference in length between the two paths he takes?

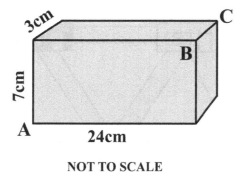

NOT TO SCALE

Figure 3.3   The closed cardboard box.

3.4 Where there's a will

Every afternoon, Wandering Will leaves his school (white) to walk to his home (black). His two grandmothers live in houses (grey) on possible routes along the way, one to each house. Now, he always visits one of them for tea and cake, but never both of them during the same journey home. The same road (grey) is never traversed more than once, and Will never visits the same road junction more than once. Not all roads are necessarily used in a single journey, and once he reaches home his journey ends.

How many different routes can he choose from?

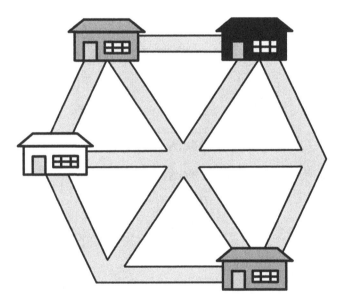

Figure 3.4   Will's home, school, and two grandmothers.

3.5 Cutting the carpet

Antwit wishes to make a new carpet from a rectangular one that measures 8m x 4m (see Figure 3.5). He folds the carpet along the dotted line so that point A rests on point C and then he flattens the fold. Any carpet that is one-layer thick is trimmed off with cutters. When the remaining double-layer is unfolded, a carpet in the shape of a rhombus results.

What is the area of Antwit's new carpet?

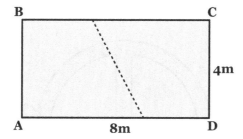

Figure 3.5   Carpet with the required fold.

3.6 Distance learning

Every afternoon, Jogger Jane runs from her home (left) to the school (right) to collect her child. Each of the four straight roads is 1 km long and each of the four curved ones is 1.5 km. She always runs more than 3 km, and in doing so, she never passes along the same road twice. Not all roads are necessarily used in a single run, she can pass by her home, and once she reaches the school her run ends.

How many different routes can she choose from?

Figure 3.6   Possible routes from Jane's home to the school.

3.7 Flea bytes

At Eccentronics Ltd, the latest robotic flea is being tested in the laboratory. A numbered 3×3 grid of squares is marked out (see Figure 3.7) and a number of trials are performed.

A trial consists of the following. A digit from 1 to 9 inclusive is randomly selected, and the flea is then placed on that grid square. The flea has been programmed to jump to an adjacent square either horizontally or vertically which it selects at random. It can revisit squares. After three jumps, its final position is recorded. The test stops when the total number of trials in which the robotic flea has finished in the middle square reaches four.

What is the expected total number of trials?

Figure 3.7 The robotic flea's grid.

3.8 Delivery dilemma

Bob the newspaper boy is about to begin his new evening delivery round. Starting at the newsagents at A, he must visit each house once only, moving one place at a time either vertically or horizontally to an adjacent house, until he reaches his own home at B. His intention is to take a different path every day until he has covered all possible routes that pass through all the houses.

How many days will it take him?

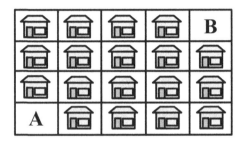

Figure 3.8   Delivery round from the newsagent to Bob's home.

3.9 Heightened awareness

Sam Trout was showing his friends his new fish tank. When he placed a brick on the bottom of the tank resting upright, the water level rose to 4/5 of the tank height. After similarly adding an additional identical brick, the water rose an additional 2/15 of the tank height.

The tank originally contained only water and the height of an upright brick was greater than the tank height. The bricks and the inside of the tank were shaped like rectangular boxes.

What fraction of the tank height was the original water level?

3.10 Handel's aria

Court composer George Frederic Handel was becoming anxious. The royal court had commissioned a new opera "Lifetime in Lockdown," but he was having trouble following their guidelines. It was no easy task when the notes had to be two metres apart. So Handel decided to get a grip on himself. To sharpen his focus and socially distance himself from his household, he built a secluded area in his garden. The retreat consists of a rectangular lawn surrounded on three sides by a fence (dark grey) and on the fourth side by a wall (white).

The enclosure has area $(x + 23)$, with dimensions $(y + 1)$ and z. The values x, y, z are whole numbers greater than one, and their total is equal to the length of the fence (which we assume to have negligible thickness). The digital sum of the three numbers x, y, z equals the sum of the digits in their total. (For example, if $x = 27$, $y = 17$, $z = 8$ then their digital sum is $2 + 7 + 1 + 7 + 8 = 25$.)

What is the area of the enclosure?

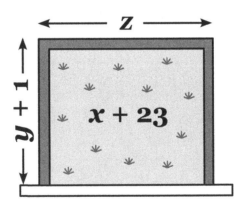

Figure 3.9   Handel's secluded area in his garden.

3.11 Gimme shelter!

Keith Jagger wants to build a nuclear shelter in the Wood at the back of his house. This is to be made from a cubic number of cube-shaped Stones. However, to avoid getting no satisfaction, he needs to make sure he orders the right number.

The shelter will have the external dimensions of a cube (see Figure 3.10). The four walls and roof are each to be one Stone thick, but the base will have no Stones. The whole structure will be built on a concrete platform. A space will be left in one wall to accommodate a steel door. The size of the door space in number of Stones is to be the same as the length (or width) in Stones of the shelter interior. Needless to say, on completion Keith Jagger intends to Paint It Black!

If all the Stones are to be used, Watts the smallest number that could be ordered?

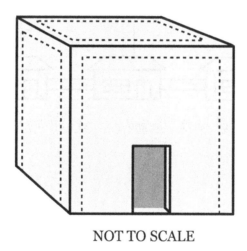

NOT TO SCALE

Figure 3.10   The cube-shaped nuclear shelter.

3.12 Santa's warehouse

Santa needs to travel from his warehouse (white) to his grotto (black), by passing through his workshops (grey), but without passing along the same path (grey) more than once. Not every workshop need be visited on his way to the grotto.

In how many ways is this possible?

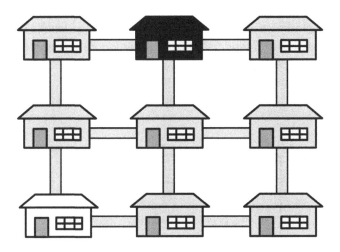

Figure 3.11   The grid of Santa's workshops.

3.13 Toby's tank teaser

Toby the toddler has seven identical cuboid-shaped sticks. The width of each equals its depth while its length is longer than the width. In the living room is a cuboid-shaped tank, the inside base of which measures 3 × 1 stick lengths. It contains a volume of water. Toby places one of his sticks length-ways in the water, so that it rests on the bottom of the tank (see Figure 3.12). The water level just reaches the top face of the stick. Then he removes it and stands all seven sticks upright in the water, so that they rest on the bottom of the tank. The level again reaches the top faces of the sticks.

What number of sticks is equal to the volume of water in the tank?

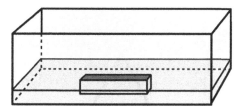

Figure 3.12   Stick lying down in the water.

3.14 Thinking in parallel

Leading onto Professor Neuron's garden is a large triangular stone patio constructed from 16 smaller triangular paving stones. In order to plant the seeds
of good health, he decides to remove a number of these stones and sow a
section of lawn in the shape of a parallelogram. So in a fit of benevolence,
he employs his nephew Trifle to take on the task. However, the teenager has
realised that there are many sizes and orientations of parallelogram from
which a lawn might be made (one of which is shown above in dark grey in
Figure 3.13).

A parallelogram is a four-sided shape with opposite sides that are both
parallel and equal length. It is possible, but not necessary, for adjacent sides
to have equal length.

How many different parallelogram-shaped lawns are possible?

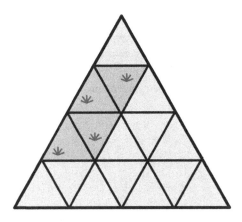

Figure 3.13   A possible parallelogram lawn.

3.15 Pentomime

At the Battle of Bosworth field in 1485, Henry Tudor was sitting in his tent admiring his set of 12 pentominoes. A pentomino is an arrangement of five identical squares joined edge to edge, and there are 12 possible shapes. Henry had asked his court mathematician to fit them all together on a 6 × 10 marble base as shown.

Suddenly, a lead cannonball whistles through the tent and shatters the marble base into four pieces, as indicated by the dotted lines above. In the process, two of the pentominoes (shown black in Figure 3.14) are destroyed, but the other ten manage to survive intact. Two of the four marble fragments that remain are 5 × 5 squares, and one pentomino in each section (shown dark grey) is still in position from the original arrangement.

How can the remaining eight pentominoes (shown light grey in the top rectangle) be fitted into the two squares?

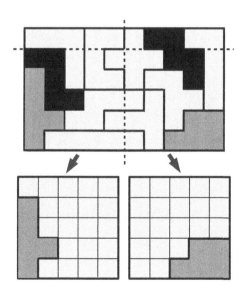

Figure 3.14   The division of the rectangle into two smaller squares.

Solutions

3.1 Calculating feat

The diameter of the lawn is 26 metres. Let the circle centre be O. After joining OC, OCB is a right angle and the radius OA = OC. Let OB = a, OC = b, and BC = c. Then $a + b = 14 \times 7 = 98$. Let $c = 7n$, where $n < 14$ is a whole number of paces along BC. Using Pythagoras' theorem, since $a^2 - b^2 = (7n)^2 = (a + b)(a - b)$ then $a - b = (7n)^2/98 = n^2/2$. If the lawn radius b is a whole (prime) number then given $a + b = 98$, a must also be whole, $n^2/2$ is integer, and so n can only be an even number. Values for a and b are obtained using their sum and difference at various even $n < 14$. Only for $n = 12$, $a = 85$, $b = 13$, $c = 84$, is the radius b a prime number. The diameter is twice the radius b.

3.2 Paper trial

There are 24 possible routes.

Double-arrow means direction has priority.

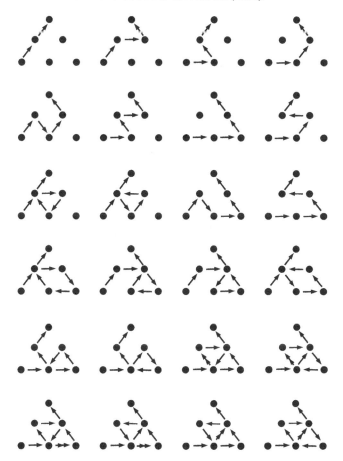

Figure 3.15 The possible routes.

3.3 The bee and the box

The difference is $26 - 25 = 1$ cm. The path from A to B in the closed box is calculated from a two-dimensional Pythagoras' theorem, so $\sqrt{7^2 + 24^2} = 25$. Likewise, the path from A to C is $\sqrt{(7+3)^2 + 24^2} = 26$.

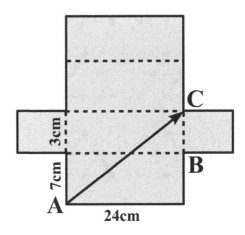

Figure 3.16   The net for the closed box problem.

3.4 Where there's a will

There are 12 possible routes.

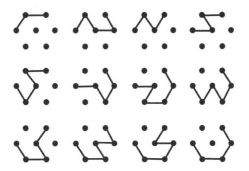

Figure 3.17   The possible routes.

3.5 Cutting the carpet

Its area is 20 m². The area of the rectangle is 8 × 4 = 32. Let the hypotenuse of the lower triangle formed after folding be x (see Figure 3.18). Then the base must be $(8 - x)$. Using Pythagoras' theorem gives $x^2 = (8 - x)^2 + 4^2$, which has the solution $x = 5$. So the area of the two triangles cut away is 2 × $(8 - 5) × 4/2 = 12$. Then 12 subtracted from 32 gives 20.

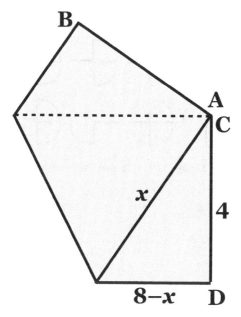

Figure 3.18   The folded carpet makes a triangle in the bottom right corner.

3.6 Distance learning

There are 16 possible routes that exceed 3 km.

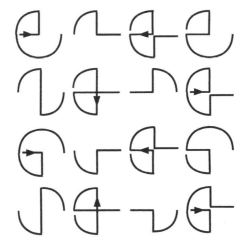

Figure 3.19   The possible routes.

3.7 Flea bytes

On average, 27 trials are required.

The problem can be reduced to three cases according to the position of the initial grid square: a corner (1, 3, 7, 9), a middle of side (2, 4, 6, 8), or the centre (5). For each of the nine positions in the grid, the possible next visits can be listed as follows: (For square number 1) followed by 2 or 4; (2) 1, 3 or 5; (3) 2 or 6; (4) 1, 5 or 7; (5) 2, 4, 6 or 8; (6) 3, 5 or 9; (7) 4 or 8; (8) 5, 7 or 9; (9) 6 or 8. However, with three jumps it is only possible to end at the centre by starting at a side. The possible combinations starting at 2 are (a) 2125, 2145, 2325, 2365, and (b) 2525, 2545, 2565, 2585. We can draw a probability tree diagram showing the possible routes when starting from 2.

The total probability of reaching a 5 for group (a) is $(1/9)\times(1/3)\times(1/2)\times(1/3)\times4$, and for group (b) is $(1/9)\times(1/3)\times(1/4)\times(1/3)\times4$. The sum is 1/27. With four possible middle-of-side starts, the total is 4/27.

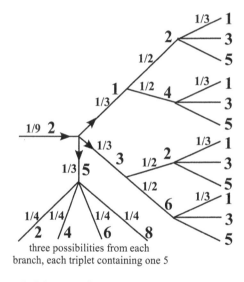

three possibilities from each
branch, each triplet containing one 5

Figure 3.20   Probability tree diagram starting from 2, the middle of a side, and ending at 5.

3.8 Delivery dilemma

It will take Bob 20 days as there are 20 different routes.

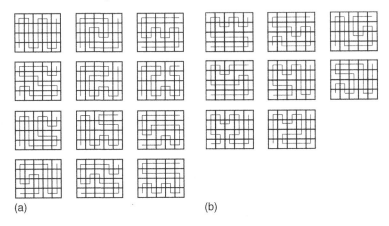

(a)   (b)

Figure 3.21   Possible routes for Delivery dilemma.

3.9 Heightened awareness

The original water level was 7/10 of the tank height. Let the original fraction of the tank height be x, the inside area of the base of the tank be A, and the area of an upright brick in contact with the inside tank base be B. The area of water in contact with the bottom of the tank when a single brick is introduced is $(A - B)$, and becomes $(A - 2B)$ for two bricks. So since the water volume is the same when no bricks, one brick, and two bricks are in the tank we have

$Ax = (A - B)4/5 = (A - 2B)14/15$. This gives $A = 8B$ and $x = 7/10$.

3.10 Handel's aria

The area is 36. The fence equation is $x + y + z = 2y + 2 + z$ from which $x = y + 2$. The area equation is $z = (x + 23)/(y + 1)$ which after substituting x becomes $z = (y + 25)/(y + 1)$. Running up through the y starting at $y = 2$ and looking for a whole number z, the only possibilities are $(x, y, z) = (4, 2, 9), (5, 3, 7), (7, 5, 5), (9, 7, 4), (13, 11, 3)$, or $(25, 23, 2)$. Their totals are 15, 15, 17, 20, 27, and 50, respectively. The triplet $(13, 11, 3)$ has a total of 27, and so both of the digital sums is 9. This gives $x = 13$ and the area is $13 + 23 = 36$.

3.11 Gimme shelter!

Keith Jagger needs 512 Stones. Let the external dimension of the cubic shelter be n. Then he needs a cubic number N consisting of the filled in solid shelter n^3; minus the Stones in the internal cube $(n-2)^3$ with the walls, roof, and base one Stone thick; minus the internal base $(n-2)^2$ one Stone thick; and minus the door space $(n-2)$. So $N = 5n^2 - 9n + 6$. Running up the $n > 2$, the smallest value for which N is a cube occurs at $n = 11$. Then $N = 512$. This puzzle is my tribute to the Rolling Stones song "Gimme Shelter."

3.12 Santa's warehouse

There are 12 possible routes.

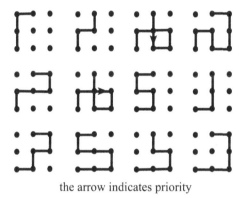

the arrow indicates priority

Figure 3.22   The possible routes.

3.13 Toby's tank teaser

The water has a volume V equal to five sticks. Let a stick have length x and base area y^2. In each scenario, the water volume V is the tank base area $3x^2$ times the water height (y or x), minus the total volume of the submerged sticks. So $V = 3x^2y - xy^2 = 3x^3 - 7xy^2$. So $2y^2 + xy - x^2 = (2y - x)(y + x) = 0$. The only solution is $y = x/2$ which means a stick has volume $xy^2 = x^3/4$ and the water has volume $5x^3/4$, five times the stick volume.

3.14 Thinking in parallel

There are 45 possible parallelograms in a main triangle of side length 4. A useful technique for systematic counting would be to divide the parallelograms into their four possible types (1 length × 1 length, 2 × 1, 3 × 1, 2 × 2) and perform a count for each. These numbers are 18, 18, 6, and 3, respectively. It is possible to find the general formula for such a count. For a main triangle of side length n, the total number of parallelograms of any size is given by $n(n^2 - 1)(n + 2)/8$. In the present case, we have $n = 4$ and $(4 \times 15 \times 6)/8 = 45$.

3.15 Pentomime

In April 1994, I wrote a computer program to find all the ways 5 out of 12 pentominoes could fit into a 5 × 5 square. This was after rotations and reflections had been disregarded. My program found 102 solutions, and out of them grew this puzzle which is about the number of ways 10 out of 12 pentominoes can occupy two 5 × 5 squares. For this, I found eight possible pairs of squares.

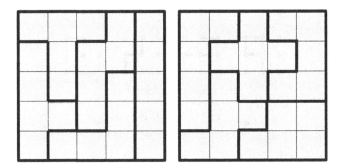

Figure 3.23   The pentomino squares.

Arithmetic

4.1 Grunting Grog

The barman of the Pig and Bucket receives an order from three separate tables for a number of pints of Grunting Grog. These numbers are 7 pints, 10 pints, and 16 pints of beer. However, when he puts the pints on the bar he divides them into three equal groups. When a representative from each table comes to collect their order, they realise that the required three numbers can be produced by moving bottles between the three groups. They move 3 pints from one group to another, then 2 pints from one to another, and finally 1 pint. Each of the three groups is involved in at least two transfer operations (where a single transfer operation involves only two groups) so that each time there is either a gain or loss of pints in each group of the cooperating pair.

What operations are required?

DOI: 10.1201/9781003358275-4

4.2 Balancing act

At Jack the grocer's market stall, a system of scales of negligible weight is balanced by three types of fruit: strawberries, pears, and pineapples. However, each of the single fruit at A, B, C, and D is covered by a cloth so their identity is unknown.

What are the fruits at A, B, C, and D?

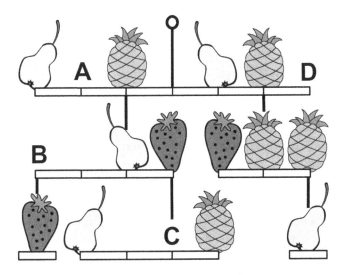

Figure 4.1   The grocer's fruit balance.

4.3 Cannibals

The Chompem cannibals of Drybone Island have the annoying habit of eating each other. In fact, Grandpa Chompem went missing only last week and it was thought he'd lost his way home until someone found his bits and pieces lying on the beach.

Now, one evening the Chompems threw a dinner party, one of those bring-your-own-food affairs—the invitations read 'bring a friend'. Six cannibals turned up, and they decided to eat each person in turn. So someone was selected for the others to eat (except the victim!), and when that person had been eaten, someone else was selected, and so on. Usually, it took each cannibal alone, two hours to eat just one person. We can assume that the six cannibals are initially identical in every respect.

How long was it before just one consumer remained?

4.4 The green gardener

Sam the handyman is having trouble getting grass to grow on his lawn so has decided to paint it green. In his shed are three paint pots containing exclusively green, yellow, and red paint. Their capacities when full are 1, 2, and 3 litres, respectively, each pot being exactly half full.

In order to obtain the right shade of green, Sam needs to mix the paints, so he pours half of the green paint into the yellow pot and thoroughly mixes them. Half of this mixture is then introduced to the red pot and likewise thoroughly stirred. Finally, the mixture from the red pot is tipped back into the green pot until it is once again half full. No paint is wasted in the process.

How many litres of green, yellow, and red paint are now in the green pot?

4.5 Sweet solution

Professor Neuron was stuck on a rather sticky problem. Three boxes A, B, and C each contain a number of chocolates. There are three different types of numbers as follows:

(1) an odd number that is not square;
(2) an even number that is not square;
(3) a square number that can be odd or even.

Initially, box A contains type (1), B has type (2), and C holds type (3). Three chocolates are now moved from B to A, then five from A to C, and then four from C to B. Before and after each move, the following two conditions must be satisfied:

(a) there is always one type of number in each box with no two boxes containing the same type;
(b) at any stage, each box can only contain a number from 1 to 9 inclusive.

How many chocolates were initially in boxes A, B, and C?

4.6 Tree total

Willie the woodcutter was sitting on the mountain side, enjoying a flask of tea, and contemplating the task in hand. Three oak trees, A at 21 m high, B at 25 m, and C at 26 m, are each to be trimmed to the same height. Altogether nine different whole number lengths are to be cut off, three from each tree, and the pieces are to be removed in the order 9 m long, 8 m, 7 m, and so on down to 1 m. No tree is to receive two consecutive cuts.

Willie was scratching his head as to how to proceed. Suddenly, he'd twigged it! He threw his flask aside, fired up his chainsaw, and immediately set to work. When he had finished, although he was satisfied with his work, he noticed that his final cut was not OK (cryptic clue)!

In what order of oak trees were the nine cuts made?

4.7 A very open prison

Dodger Dave has just picked the lock on his prison cell and now intends to make his escape by running down a long straight corridor. The corridor is partitioned by three doors A, B, and C, that open at regular intervals. Door A opens every 10 seconds, B every 25 seconds, and C every 15 seconds, each door staying open just long enough to pass through.

Dave plans to make his run at a constant speed in a straight line without stopping. He knows that it will take 25 seconds to reach door A from his cell, 10 seconds to run from A to B, and then 15 seconds from B to C. Suddenly, he notices that the three doors have opened simultaneously. He now wants to make his escape in the shortest possible time.

How many seconds should he wait before beginning his run?

4.8 Oh deer!

Down at the grotto, Santa intends to pose Rudolph a puzzle. Before he arrives, Santa repeatedly throws a traditional dice until the last four throws are such that no two of the four results are identical. This is the four-digit number that Rudolph must deduce.

As Rudolph trots in, Santa gives him four statements about the number:

(1) the first digit is odd,
(2) the second is prime,
(3) the third is even,
(4) the fourth is square.

However, something is amiss. Dasher, who was present during the throws, waits until Santa takes a break then informs Rudolph that exactly two of these four statements are lies. He then gives the following reliable facts about the number:

(a) The second and third digits differ by 3.
(b) The first and fourth total 7.
(c) The sum of the four digits is 12.

What was Santa's four-digit number?

4.9 See-saw sums

Five friends Fred, George, Harriet, Ivy, and John have masses 37 kg, 74 kg, 111 kg, 148 kg, and 185 kg, respectively. They climb onto a wooden see-saw (see Figure 4.2) with one to a seat A–E and arrange themselves so that it balances. One person now climbs off on one side, the middle person remains seated, and the remaining three rearrange themselves, one to a seat, so that the see-saw balances again.

The seats are equally spaced and the second seating arrangement bends the see-saw more than the first.

Who sat in the middle and who climbed off?

Figure 4.2   The see-saw.

4.10 Guarding the gold

Down at Whimsy Woods, the pixies have been alerted that a wicked elf is planning to steal their biggest sack of gold. The pixies keep their savings in four tree trunks A–D which each contains a single sack of gold coins. The sack in tree A contains 1 kg, B has 2 kg, C contains 3 kg, and D has 4 kg. So to confound the elf, they have decided to changes the bags around.

Three sacks are selected from the four to participate in three consecutive moves, no sack to be moved more than once. In turn, a chosen sack is carried by the pixies to one of the other three trees. After the three moves, there is once again a single sack of coins in each tree but they now appear in a different order. Given that all amounts are in kilograms (kg), the final order can be deduced from the following facts.

(1) The first amount moved was greater than the second.
(2) After the second move, the total amount in C was greater than the sum in the other three trees.
(3) After the third move, the total amount in B increased.

Can you find the amount of gold coins in each tree after the three moves?

4.11 Beetle mania

The Black Clock Beetle lives under a stone and according to the observations of that famous evolutionary biologist Dr Dee Fax, it usually makes an appearance at regular intervals during the hours of darkness. Last Sunday as her pocket watch struck midnight, Dr Fax saw three of them emerge together from under a large stone. However, no sooner had they appeared, they quickly returned to cover. It turned out that the first beetle appeared every 84 seconds, the second every 90 seconds, and the third every 98 seconds.

Now Dr Fax wanted a photograph in which all three beetles appeared at the same time. Unfortunately, her mounted flash camera was only primed to take a picture on the hour every hour, but at midnight she discovered that she had forgotten to remove the lens cap. Nevertheless, she knew there had to be a moment in which the appearance of all three beetles again occurred simultaneously.

How many hours after Sunday midnight did it take to obtain the required photograph?

4.12 Counting sheep

Above is an 8 × 8 grid of sheep pens with the number of sheep given in each pen. Farmer Jim has found that it is possible to delete two numbers from each row and column, leaving 48 numbers, so that each row and column totals 30.

Can you reproduce the grid after the deletions have been made?

8	5	9	1	9	5	9	1
5	9	6	8	7	3	4	5
7	2	8	3	6	5	7	4
4	6	1	6	3	7	7	9
9	7	4	5	4	3	6	6
5	5	7	9	5	9	1	5
6	8	2	4	3	5	5	8
3	3	4	7	5	7	5	4

Figure 4.3   Number of sheep in each of 64 pens.

4.13 Turf teaser

Gordon the gardener has been hired to re-turf Lord Klump's lawn, so he removes an area A of the old turf to leave an area B which is smaller than A (see Figure 4.4). Gordon has discovered that the three areas—the whole lawn, A, and B—are each the square of a positive integer number of metres, no two numbers being identical, and exactly two of these positive integers are two-digit prime numbers. Knowing that Lord Klump has the smallest lawn possible under these conditions, Gordon buys a number of rolls of turf each having the same area as B.

What is the smallest whole number of rolls needed to ensure area A is re-turfed?

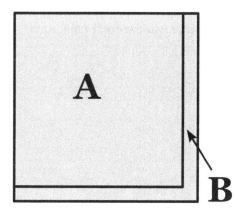

Figure 4.4   Lord Klump's lawn.

4.14 Taking the pizza

Five friends were debating whether or not they should buy a pizza to share between them. So everyone emptied their pockets. It turned out that Agatha had £1, Babble had £2, Crumble had £3, Dibdib had £4, and Earwig had £5. Each then claimed that some combination of two of them had the exact total money to pay for the pizza.

(1) Agatha said "me and 'Crumble or Dibdib'"
(2) Babble claimed "me and 'Dibdib or Earwig'"
(3) Crumble stated "me and 'Babble or Dibdib'"
(4) Dibdib said "me and 'Agatha or Earwig'"
(5) Earwig stated "me and 'Agatha or Crumble'"

Exactly three of them were lying and the 'or' is mutually exclusive.

Who were the liars and what was the cost of the pizza?

4.15 Good health!

A good elf was walking through Whimsy Wood carrying a tray of seven full and seven empty champagne glasses, to visit two of his friends. Suddenly, he came across a circle of eight stones (labelled A–H in Figure 4.5). As he approached them, a fairy appeared and informed him that to continue on his journey he must climb onto one of the stones then make three jumps clockwise. In any order, two of the jumps must be to an adjacent stone and the other must omit a stone to land on one two away. Each of the four stones he touches will either empty a number of glasses that are full (negative number) or fill a number of glasses that are empty (positive number). The elf leaves with 10 full glasses.

What is the letter sequence for the four stones visited?

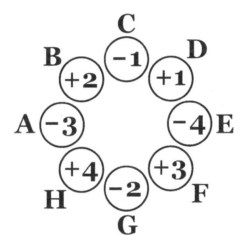

Figure 4.5   Eight stones with their effect on the glasses.

4.16 Magic multiples

Winnie the witch has just decorated her kitchen floor with a 3 × 3 grid of tiles. Each of the nine tiles has a number of broomsticks in its pattern, a single digit from 1 to 9, where no two digits are identical. As might be expected, the numbers form a 3 × 3 magic square, where each row and column has a total of 15. In contrast to a standard magic square, this is not necessarily the total for the two diagonals. This magic square above has some curious properties.

(1) The digit G is a multiple of H.
(2) The sum of the three digits F, H, and I is a multiple of A.
(3) The total of the nine digits is a multiple of the sum of the digits ABD.

Can you fill in the nine digits A–I?

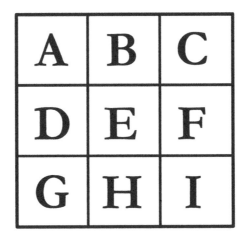

Figure 4.6   The nine tiles.

4.17 The puzzling page

Professor Neuron's nephew Trifle was proud to have his first mathematical puzzle published in *Reasoning Weekly*.

"All I can remember," said Trifle, "is that it's a two-digit number, neither digit being zero, and less than ten pages before the middle of the publication."

"Interesting," said Neuron.

"Also, if one reverses the two digits and counts this new number of pages from the back page towards the front, one arrives at the puzzle page."

Neuron knew that this particular magazine always had a multiple of 15 pages. His curiosity was now roused. "And what's the puzzle about?"

"Well," said Trifle, "it's about an absent-minded young fellow who publishes a mathematical puzzle in a magazine but can't remember ..."

What page was Trifle's puzzle on?

4.18 Digital deletions

At Perspica City University, Professor Neuron was taking a class of mathematics students. As he prowled around the class checking his students' work he discovered that Wong Sum was doing work of his own.

"It's a new kind of subtraction," said Wong. Neuron wasn't convinced. "I'll show you," said Wong.

All you have to do is delete one digit from each row to leave three columns of digits—and it need not be the same position in each row, rub out a second digit from each row to leave two columns, and finally a third to leave one column, closing the gaps after each deletion. What you'll find is that after each set of seven eliminations the sum still works.

Reading down the columns, what is the third set of seven deletions?

$$
\begin{array}{r}
4\ 6\ 5\ 7 \\
+\ 2\ 6\ 3\ 4 \\
\hline
7\ 2\ 9\ 1 \\
-\ 2\ 4\ 3\ 5 \\
\hline
4\ 8\ 5\ 6 \\
+\ 2\ 7\ 8\ 9 \\
\hline
7\ 6\ 4\ 5
\end{array}
$$

Figure 4.7   The curious arithmetic problem.

4.19 Adder revelation

Mr and Mrs Sum were sitting on the patio, chatting over a pot of tea, and watching their three children Gillian, Bertrand, and Philippina playing in the back garden.

"GBP," said Mrs Sum.

"What's that?" asked Mr Sum.

"Their initials are GBP, and the total money they have in their piggy banks is 16 GBP."

"Exactly," said Mr Sum, "and although each of their amounts is a whole number, no two amounts are the same."

"And don't forget that only one of their amounts is a prime number," said Mrs Sum.

Mr Sum took another sip of his Camomile tea.

"Only one child's amount is the number of letters in their name. Isn't that curious?"

"Yes," said Mrs Sum, "and of the other two, only one of them has an amount one less than their number of letters."

"Don't you think this is a strange conversation?" said Mr Sum, stirring the tea with one of his digits.

"Yes," said Mrs Sum. "Maybe we should get out more."

What amounts belong to Gillian, Bertrand, and Philippina?

4.20 Cryptic cards

Professor Neuron was tormenting his nephew Trifle with a card trick. He placed five cards in a line on the table (see Figure 4.8), three face down, with the jack of diamonds and the five of spades face up. Neuron then announced that the following facts are true about the five cards.

(1) There is at least one card of each suit, but no two cards of the same suit are adjacent.
(2) Exactly two of the face down cards total 20.
(3) No club is less than 6.
(4) There are two hearts, exactly one of which is a court card.
(5) The last three cards total at most 13.

An ace has the value 1, a jack is 11, a queen is 12, and a king is 13.

Can you identify the face down cards from left to right?

Figure 4.8   Neuron's card trick.

4.21 Socks and shares

On Planet Pogo, the Unipeds manage to attend social events by hopping around on their only leg. At one of their sock parties, five boxes are produced each containing the same number of socks. Father Hop takes three boxes and his wife Mother Hop takes the remaining two so that they each have less than 100 socks each.

"That's not fair," says Mother. "You have more socks than me."

"Darn it!" says Father, and with that he gives one sock to his wife.

"But we don't have any!" exclaims Skip, their eldest child.

So in a fit of benevolence, Mother now shares all the socks in her boxes equally between herself, Father Hop and their three children. However, Father Hop does not add them to his boxes. At this point, three of their neighbours call in, so Father distributes all the socks in his boxes equally between himself, the rest of his family, and their three visitors.

How many socks are originally in each box?

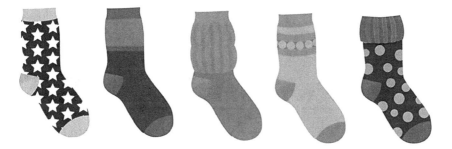

Figure 4.9   Socks from the sock party.

4.22 Fruit and nut

At the Fifteen Fruit Factory, they package their oranges in a set of three triangular boxes. Each box in the set has a white, grey, and black compartment containing a number of oranges (see Figure 4.10). Not only does each box contain 15 oranges, but the total for each of the three colours is also 15.

One night, a malevolent intruder enters the premises and tampers with the boxes. He makes three exchanges so that the above conditions are still satisfied. Each exchange involves two different boxes of the three, so that all the oranges in the compartment of one box are juxtaposed with all those in a compartment of another. Altogether, six compartments are involved, no two being the same, and each box participates in two exchanges. It turns out that the total of the three compartments *not* involved is also 15. Also, for any two numbers involved in a juxtaposition, their sum is no greater than 12.

What are the three pairs of numbers involved in the exchanges?

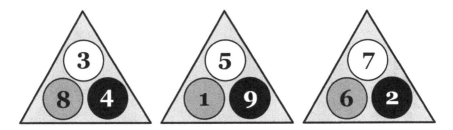

Figure 4.10   Number of oranges in each circular compartment of a box.

4.23 Lettuce play

Rabbit has forgotten to hand in his mathematics homework. Unfortunately, his mathematics master, Mr Warren, is a religious fanatic and believes that sins should not go unpunished. So, by way of retribution, he has secretly changed the two-digit pass number on Rabbit's school lunch box. When the bell rings for lunch, and Rabbit removes the box from his satchel, he is disappointed to discover that he is unable to eat his lettuce sandwiches.

Realising what has happened, Rabbit asks his teacher for the new pass number, but instead he is given seven statements, exactly three of which are false.

(1) The two digits sum to 9.
(2) Exactly one digit is a square number.
(3) The two digits sum to 10.
(4) Both digits are odd numbers.
(5) Exactly one digit is a prime number.
(6) Exactly one digit is divisible by 3.
(7) The two digits sum to 11.

The two digits in the pass number are 1–9 inclusive, no two digits being identical, and the higher digit precedes the lower one.

Mr Warren believes it would take a miracle for Rabbit to unravel the enigma. However, he has underestimated his ability. Rabbit leaps on the problem, and with religious application soon solves it.

What is the pass number for the school lunch box?

4.24 Going bananas

Deep in the Booliba jungle, the monkeys are resting after eating too many bananas. However, they are a crazy bunch, and one of them has left some banana skins at positions A and B on the top beam. Consequently, the monkeys in those positions are about to slip off and bite the dust.

There are three different kinds of monkey: light grey, white, and dark grey. Monkeys of the same colour have an identical weight, but no two different coloured monkeys weigh the same. Our furry friends are shown arranged on hanging beams that have negligible weight, with two separate hanging beams below, and with all three beams balanced. Each monkey sits at one of the equally spaced marks. This means that if there are only two monkeys on a beam, and they have a different colour, a monkey that sits one space from a hanging point has twice the weight of one that sits two spaces away.

What colours are the two disappearing monkeys at A and B?

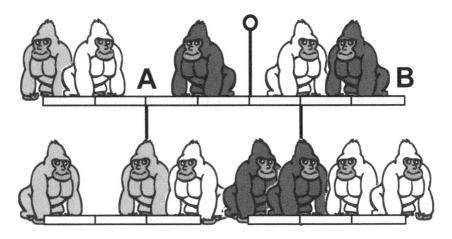

Figure 4.11   Monkeys on a set of connected balances.

4.25 No pain no gain

In the east end of London, the Kleaning-Right-And-Proper company are pay-ing five window cleaners to take their buckets and sponges to an office block. The company has issued five identical ladders each with nine rungs which are being used simultaneously by the five cleaners, one to each lad-der. When it's time to break for lunch, although each cleaner is up a ladder, no two are standing on the same number rung.

(1) Earwig's rung is one higher than Antwit's.
(2) The total of Crumble and Dibdib's positions gives the rung that Babble is on.
(3) Dibdib is two rungs below Antwit.
(4) Babble and Dibdib are one rung apart.
(5) Only one of the rung positions is a prime number.

Can you give the rung position for each window cleaner?

4.26 Milky weigh

At the Colossal Coin Company, nine different milk chocolate coins with whole number weights from 1 to 9 kg are to be selectively placed on weighing scales where the possible coin placements A–G are equally spaced (see Figure 4.12). One of these nine coins is chosen and placed at A so that when two other coins are placed together at E, F, or G, then the scales balance. If the coin on the left of the scales is now moved to B, then the apparatus will still balance if the two coins on the right from the first weighing are now separated and placed at two of E, F, and G. No two coins of the three are the same during a weighing. If no coin is present a weight of zero is to be recorded.

What are the weights of the coins at B, E, F, and G for the second weighing?

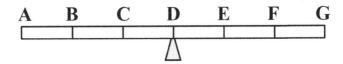

Figure 4.12   Weighing scales for chocolate coins.

4.27 Bird brain

Professor Neuron has taken his nephew Trifle to the zoo. As they reach one of the cages in the aviary, six different birds are seen balanced on perches as shown: A, B, C, D, E, and F. According to the guide book, the weight of each bird is a single digit from 2 to 9 inclusive, no digit being repeated.

"It's amazing that they manage to balance," said Trifle.

"But don't you see?" said Neuron. "It's actually possible to calculate the weight of each bird."

Evidently, two of the eight digits are left unemployed in the calculation.

Can you give the weight of each bird shown on the perches?

Figure 4.13   Perches with the six birds.

4.28 Three shades of grey

Antwit, Babble, and Crumble have just ordered the triple grey: three drinks that are light grey, white, and dark grey. Their glasses have capacity 2, 3, and 4 units, however, no glass is full and the contents are actually 1, 2, and 3 units, respectively. The 1 unit of light grey and the 2 units of white are fruit juices, but the 3 units of dark grey have a high alcohol content. When his friends visit the toilet, Crumble decides to play a trick on them by making the following pourings.

(1) The white glass is used to top up the light grey leaving 2, 1, and 3 units.
(2) The light grey tops up the dark grey leaving 1, 1, and 4.
(3) The dark grey tops up the white leaving 1, 3, and 2.
(4) The white tops up the light grey leaving 2, 2, and 2.
(5) Half of the light grey glass is poured into the dark grey restoring the 1, 2, and 3.

The contents are thoroughly mixed after each pouring. Since the original quantities are still the same, Crumble believes that his prank will pass unnoticed; however, after drinking all evening his power of perception starts to become a grey area.

What is the final ratio of the units of dark-grey liquid between the three glasses?

Figure 4.14   The three glasses.

4.29 The odd getting even

In the village of Little Point, there are six houses in a row numbered 1 to 6, no two digits being identical. Ever since the houses were built, members of the Odd family have inhabited 1, 3, and 5, while members of the Even family have lived in 2, 4, and 6. However, the families have always been in conflict over whether the six-digit number given by the row of houses should be an odd or an even number. Despite the bickering, the houses still have their original numbering from left to right as 123456 which, of course, makes an even number.

Tension has been mounting recently and renewed threats have been made by the Odds to alter the order of the digits. Locals have done their best to keep the peace. Against all Odds, Police Constable Nickham has been persuaded to intervene and every night he has been commuting by bicycle from the next village, Reese-on-Wye, on the alert for prime suspects.

Then one night, Sam Watt-Odd, hides behind a bush, waits for Nickham to complete his round, then makes his move. He changes the digital signs around to render the resulting six-digit number odd. As a result, no two prime digits end up on adjacent houses. Needless to say, the Evens wake up the next morning and find the new six-digit arrangement rather odd.

How many possible odd six-digit number arrangements could Sam have chosen from?

Solutions

4.1 Grunting Grog

Let the group that is to contain 7 bottles be A, 10 bottles be B, and 16 bottles be C. Then we have 3 from A to C, 2 from B to C, and 1 from A to B. Group A must lose 4 bottles, B loses 1 bottle, and C gains 5 bottles. Each group is involved in exactly two transfers. So C must have +2 and +3. Group B can have −2 and +1 or −3 and +2. Group A can have −2 and −2 or −1 and −3. The first option for group A means that 2 bottles are exchanged twice (invalid). So group A is −1 and −3. Then group B is −2 and +1.

4.2 Balancing act

A strawberry (S) is at A, a pear (R) is at B, a pineapple (P) at C, and a pear at D. On the bottom middle scales, taking moments (weight times distance), $2 \times R = 1 \times P$, so P = 2R. On the bottom right arrangement, $1 \times S = 1 \times P + 1 \times R$, so S = 3R. On the middle left scales, denoting the missing fruits as B and C, $2 \times B + 2 \times 3R = 1 \times 3R + 1 \times R + 1 \times C + 1 \times 2R$. So 2B = C. This means B = R (pear) and C = 2R (pineapple). For the top scales we now add the weights of the lower scales to get moments. For the top left (including the B and C weights) we have $3 \times R + 2 \times A + 1 \times 2R + 1 \times 13R$ (3 pears, 2 pineapples, 2 strawberries). For the top right, we have $1 \times R + 2 \times 2R + 2 \times 8R + 3 \times D$. Equating these gives 2A = 3R+3D, so A = 3R (strawberry) and D = R (pear).

4.3 Cannibals

It took four hours and thirty-four minutes for one cannibal to be left. It took five cannibals 2/5 hours to eat the first victim, four cannibals 2/4 hours for the second, three cannibals 2/3 hours for the third, two cannibals 2/2 hours for the fourth, and one cannibal 2/1 hours for the fifth. These sum to the given solution.

One could argue that stomach contents increase a cannibal's mass when a victim. However, if the further interpretation is granted that these contents also proportionally increase a consumer's capacity to devour, then the given solution still holds! This previously published puzzle of mine [1] deserves inclusion for this added twist.

4.4 The green gardener

The green pot contains 9/34 litres of green, 1/17 litres yellow, and 3/17 litres red paint. When 1/4 litre of green paint is poured into the yellow the mixture obtained is 5/4 litres. Half of this mixture is tipped into the red pot to produce 1/8 litre green, 1/2 litre yellow, and 3/2 litre red, a total of 17/8 litres. However, only 1/4 litre, or 2/17, of this mixture is poured back into the green pot to restore it to 1/2 litre. This amounts to 2/17×1/8=1/68 litres of green paint which together with the 1/4 litre of green left in the pot totals to 9/34 litres. For the yellow we have 2/17×1/2=1/17 litres and for the red we have 2/17×3/2=3/17 litres.

4.5 Sweet solution

Box A starts with 3 chocolates, B with 6, and C with 1. Type (1) contains 3, 5, or 7; type (2) has 2, 6, or 8; and type (3) must be 1, 4, or 9. For the given conditions to be satisfied, 1 or 3 chocolates must finish in A, 7 or 9 in B, and 2 or 5 in C. In the final numbers, only C can have an even number which is 2. So, box C must consecutively contain 1, 1, 6, and 2. If the final square number is 9 in box B, then B consecutively has 8, 5, 5, and 9; so A must finish with 3 (not 1 which is square) and has 5, 8, 3, and 3. This gives no square number in any box after the move A to C. So the final square number must be 1 in A which consecutively has 3, 6, 1, and 1, leaving 7 in B and the sequence 6, 3, 3, and 7. For each box, the starting number in its sequence gives the solution.

4.6 Tree total

The trees were cut in the order CBCABABAC. The total height of the oak trees before cutting is 72 m and with the nine cuts a total of 45 m is lost. So their total height after cutting is 27 m, which when divided by three gives each as 9 m high. This means A loses 12, B loses 16, and C loses 17. With no tree receiving two consecutive numbers, and the numbers diminishing, the possibilities for A are 831, 741, and 642; for B we have 961, 952, 862, 853; and for C there is 971, 962, 953, and 863. To use up all nine cuts, no digit being repeated, the possible combinations are A = 741, B = 952, C = 863; A = 741, B = 862, C = 953; A = 741, B = 853, C = 962; and A = 642, B = 853, C = 971. However, if the last cut of 1 m was not OK (cryptic clue: oak A!), then there only remains A = 642, B = 853, C = 971. So the order is C: 9 m, B: 8 m; C: 7 m, and so on as above.

4.7 A very open prison

Dave should wait 115 seconds. Let a, b, and c, be the number of times doors A, B, and C open after their simultaneous opening. The number of seconds that elapse to a door's next opening is then $10a$, $25b$, and $15c$, respectively. Dave must reach door B 10 seconds and door C 25 seconds after A opens. So the time at which A is favourably open for the run is $10a = 25b - 10 = 15c - 25$. A good place to start is to rearrange the first two parts into $2(a + 1) = 5b$. By running up the positive integers b, we have $(a, b) = (4, 2)$, $(9, 4)$, $(14, 6)$ and so on. With this we find that agreement first occurs at 140 seconds when $a = 14$, $b = 6$, and $c = 11$. Allowing for the 25 seconds to reach A, Dave should set off 115 seconds after their synchronous opening.

4.8 Oh deer!

The number is 5142. The solution proceeds by listing the possible digits for Santa's four statements when each is assumed to be true (T) then false (F).

From (b) and (c), the 2nd and 3rd digits can only sum to 5. So we focus on statements (2) and (3). From (a), we can either have the second digit false (F) with the digit as 1 (1F) and the third statement true with the digit as 4 (4T), or we can have 4F and 1F. From (b), without repetition, the 1st and 4th can be: 5T and 2F, or 2F and 5F. We aim to combine these to produce two T and two F statements. The only possibility for the four digits is then 5T, 1F, 4T, and 2F.

Table 4.1 List of possible digits for each of Santa's statements

statement	T	F
1	1, 3, 5	2, 4, 6
2	2, 3, 5	1, 4, 6
3	2, 4, 6	1, 3, 5
4	1, 4	2, 3, 5, 6

4.9 See-saw sums

Harriet (111 kg) was in the middle and Fred (37 kg) climbed off. Dividing each weight by 37 reveals that the five weights are in the ratio 1 : 2 : 3 : 4 : 5. Considering the seats from either the left or right end, there are only three possible arrangements for the balancing of weights with all five seats occupied: (a) 5, 1, 2, 3, 4 with weight 2 seated centrally and the total distance times weight on each side $(2 \times 5) + (1 \times 1) = (1 \times 3) + (2 \times 4) = 11$; (b) 4, 1, 3, 5, 2 (total 9); and (c) 2, 3, 4, 5, 1 (total 7). The only possible rearrangements for (a) are 3, _, 2, 4, 1 (5 climbs off), 1, 3 2, 5 _, (4 climbs off), or 3, 4, 2, _, 5 (1 climbs off); for (b) we can have 5, _, 3, 2, 4 (1 climbs off), 2, 1, 3, 5, _ (4 climbs off), or 1, 2, 3, 4, _ (5 climbs off); and for (c) 2, 1, 4, 5, _ (3 climbs off) or 1, 3, 4, 5, _ (2 climbs off). Only with the first case for (b) 5, _, 3, 2, 4 does the total distance times weight $(2 \times 5) = (1 \times 2) + (2 \times 4) = 10$ increase each side (from 9 to 10) to result in a greater bending. Here, Harriet $(3 \times 37 \text{ kg})$ is central throughout and Fred $(1 \times 37 \text{ kg})$ climbs off.

4.10 Guarding the gold

Tree A has 1 kg, B has 3 kg, C has 4 kg, and D has 2 kg. From condition (2), if both of the first two moves contribute to C, two empty trees result and no single third move can produce the four original amounts (kg). So only 4 kg from D to C in one of the first two moves with no removal from C for the other move satisfies (2), and from (1) the 4 kg must be the first move to give 1, 2, 7, 0 for A, B, C, D, respectively. From (2), the second move must not leave two empty trees before the third move, for this would again prohibit finishing with the four original amounts. So either 1 kg moves from A to D giving 0, 2, 7, 1 which leaves a third move of 3 kg from C to A that violates (3), or the 2 kg moves from B to D giving 1, 0, 7, 2. This leaves a third move of 3 kg from C to B (the 4 kg has already moved) which satisfies (3), and the original four amounts are restored but in a different order as required.

4.11 Beetle mania

The three beetles were photographed together 49 hours after Sunday midnight. Let the three beetles be named A, B, and C, so that using prime factors (2, 3, 5, 7, ...) , beetle A appears every $2 \times 2 \times 3 \times 7 = 84$ seconds, B every $2 \times 3 \times 3 \times 5 = 90$ seconds, and C every $2 \times 7 \times 7 = 98$ seconds. The first time after midnight that they again appear simultaneously can be obtained by finding the smallest possible multiplier (LCM) for each so that they share the same set of prime factors. For A, the 84 must be multiplied by $3 \times 5 \times 7 = 105$, for B 90 requires $2 \times 7 \times 7 = 98$, and for C 98 needs $2 \times 3 \times 3 \times 5 = 90$ to give 8820 seconds in each case. However, this does not occur on the hour. The first hourly occurrence is when the product of the set of factors is divisible by $3600 = 2 \times 2 \times 2 \times 2 \times 3 \times 3 \times 5 \times 5$. Since $8820 = 2 \times 2 \times 3 \times 3 \times 5 \times 7 \times 7$ then an extra factor of $2 \times 2 \times 5 = 20$ is required. Here $8820 \times 20 = 176400$ seconds which when divided by 3600 gives 49 hours.

4.12 Counting sheep

	5	9	1		5	9	1
5		6		7	3	4	5
7	2	8	3	6			4
4		1	6	3	7		9
	7	4		4	3	6	6
5	5		9	5		1	5
6	8	2	4		5	5	
3	3		7	5	7	5	

Figure 4.15   Completed grid.

4.13 Turf teaser

The number of rolls of turf required is 30. The problem concerns Pythagorean triplets, three positive integers that fit around a right-angled triangle, for example, $3^2 + 4^2 = 5^2$. It is possible to generate sets of three from the calculations $p^2 + q^2$, $2pq$, $p^2 - q^2$ with $q < p$, both being positive integers. The possible sets of three triangle side lengths can be systematically calculated by starting at $p = 2$, $q = 1$, and incrementally increasing p. For example, for $p = 5$ we can try $q = 1, 2, 3, 4$. The first time that a pair of two-digit prime positive integers appears is for $p = 6$, $q = 5$ giving the triangle sides as 61, 60, and 11. This means that area A must be $60^2 = 3600$ and area B must be $11^2 = 121$. Division of area A by area B shows that 30 rolls are required.

4.14 Taking the pizza

Babble, Crumble, and Earwig were lying and the pizza cost £5. If there are three liars there must be exactly two truth-tellers. The total amount cannot be £3 because no one claims that AB have combined, and this case gives five liars. Neither can it be £4 since AC appear once only in Agatha's statement leaving four liars. For £5, AD or BC are possible pairs, and only AD produces three liars as required with BC giving four liars. For £6, AE or BD are possible, and but both of these are claimed once only with a resulting four liars which is invalid. For completeness, a check shows that £7 from BE or CD have four liars each and so are invalid, £8 from CE (which only Earwig claims) has four liars, and £9 from DE (claimed by Dibdib) has four liars.

4.15 Good health!

The sequence of stones visited is GHBC, an acronymous toast to the compiler! The act of stepping on one stone and carrying out the three jumps as described spans five consecutive stones. There are eight such clockwise sequences: A–E, B–F, C–G, and so on. Totalling the numbers for the five stones in each case gives A–E (–5), B–F (+1), C–G (–3), D–H (+2), E–A (–2), F–B (+4), G–C (0), H–D (+3). In each case, omit one of the middle three stones of the five (the one that is jumped over) to examine how the total is modified. We note that a sequence total of +3 needs to result from an omission for the 7 full glasses to become 10. So we take the opposite sign of whatever needs to be added to the five-stone total. For example, for A–E (–5) we need +8 to make +3. So we look at the middle three stones B, C, or D for one that is –8. There isn't one. However, for G–C (0) we need to add +3 to make a sequence total of +3. So we need to omit a –3 which is stone A. So, the sequence of jumps is GHBC.

4.16 Magic multiples

From A–I the digits are 6, 2, 7, 1, 9, 5, 8, 4, 3. It is possible to generate all possible magic squares by rotating or reflecting the 9 × 9 grid in Figure 4.16, and by juxtaposing rows and columns of an iterated 3 × 3 square. For example, let us focus on the 3 × 3 square in the top left corner. If we juxtapose columns 2 and 3, then juxtapose rows 1 and 3, we can find the image square has been rotated through 180 degrees in columns 2–4 and rows 1–3. Alternatively, if we reflect the top-left square on its leading diagonal then juxtapose columns 1 and 2, we discover the image square has been rotated counter clockwise 90 degrees in columns 1–3 and rows 3–5.

2 7 6 2 7 6 2 7 6
9 5 1 9 5 1 9 5 1
4 3 8 4 3 8 4 3 8
2 7 6 2 7 6 2 7 6
9 5 1 9 5 1 9 5 1
4 3 8 4 3 8 4 3 8
2 7 6 2 7 6 2 7 6
9 5 1 9 5 1 9 5 1
4 3 8 4 3 8 4 3 8

Figure 4.16   An iterated three by three magic square.

Since three rows each total 15, the nine digits total 45. For condition (3), the factors of 45 that can be constructed by adding three different digits are 9 or 15. The sum A + B + D = 15 (an 'L' shape) is not possible because with A + B + C = 15 (total of a row) then C = D which is an invalid duplication. So the sum must be 9 with triplet possibilities ABD as (1, 2, 6), (1, 3, 5), or (2, 3, 4).

Let us consider condition (2). We examine the possible ABD L-shaped cluster in the grid above, keeping in view that the cluster might appear

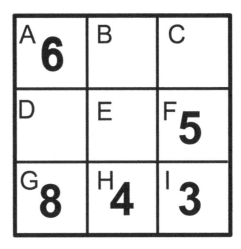

Figure 4.17   Deduction of the magic square required.

rotated. (i) For (1, 2, 6), we could have A = 6 with B/D as 1/2 (order undetermined). The opposite corner to A = 6 in the 3 × 3 square formed by this L-shape is I = 3 with F/H as 4/5. (ii) For (1, 3, 5), we can have A = 5 with B/D as 1/3 (order undetermined). The opposite corner to A = 5 is I = 2 with F/H as 4/6. (iii) For (2, 3, 4), we can have A = 4 with B/D as 2/3. The opposite corner to A = 4 is I = 1 with F/H as 5/6. Only for (i) and (iii) is (2) satisfied.

Since no digit is greater than 9, when we think about the possible values of H, condition (1) only allows case (i) with G = 8 and H = 4, so F = 5. This completely determines the 3 × 3 section of the above grid that must be extracted (see Figure 4.17). The rest follows easily from the total 15.

4.17 The puzzling page

The page number is 56. Let the two digits be x and y. The page number can then be written as $10x + y$. The reversed number is $10y + x$ but counting from the back to the page means that their sum must be one greater than the total number of pages. Also, let n be the number of sets of 15 pages. So $10x + y + 10y + x = 15n + 1$. Then $11(x + y) = 15n + 1$. The only n where $15n + 1$ is a multiple of 11 and $15n + 1 \leq 2 \times 99$ (the maximum sum of the pair of two-digit numbers) is $n = 8$ for a 120-page magazine. Then $x + y = 11$ and $(x, y) = (2, 9), (3, 8), (4, 7)$ and so on. Since the puzzle page is less than 10 pages before the middle (which is between the 60th and 61st pages) then $x = 5$ and $y = 6$.

4.18 Digital deletions

$$
\begin{array}{rrr}
4\ 5\ 7 & 4\ 5 & 5 \\
+\ 2\ 6\ 4 & +\ 2\ 6 & +\ 2 \\
\hline
7\ 2\ 1 & 7\ 1 & 7 \\
-\ 2\ 3\ 5 & -\ 2\ 3 & -\ 3 \\
\hline
4\ 8\ 6 & 4\ 8 & 4 \\
+\ 2\ 7\ 8 & +\ 2\ 8 & +\ 2 \\
\hline
7\ 6\ 4 & 7\ 6 & 6 \\
\end{array}
$$

Figure 4.18   The three stages of digital deletions.

4.19 Adder revelation

Gillian has 6, Bertrand has 8, and Philippina has 2. With a total of 16 and no two numbers equal, if exactly one of the numbers equals the number of letters in a child's name and there is only one prime number, the possible combinations are as follows: for Gillian = 7 we have (7, 6, 3), (7, 5, 4); for Bertrand = 8 there is (8, 6, 2), (8, 5, 3); and for Philippina = 10 we have (10, 4, 2). When we add the condition that only one of the other two children has a number one less than the number of letters in their name, we can only have (8, 6, 2) for Bertrand.

4.20 Cryptic cards

From (5), the third and fifth total at most 8. So the highest possible card of these two is a 7, with the other being a 1 (ace). Then from (2), a total of 20 from two of the three face down cards can only be obtained from a king (= 13) and a 7. (We cannot go lower than 13 because then we would need to go higher than 7 which is not possible.) So the second place card is a king. At this point we know the card values are 1, 5, 7, 11 (jack), 13 (king), order unimportant. From (4), the jack (= 11) is not a heart so the king must be a heart. The only problem left is to decide the order and suits of the 1 and 7 in the third and fifth positions. From (1), since a heart is second, the other heart must be fifth with a club third. Finally, from (3), the club must be the 7 in third place and so the 1 (ace) is the second heart in fifth.

Figure 4.19   The five cards.

4.21 Socks and shares

Originally there are 27 socks in each box. Let the number of socks in each box be n. So Father has $3n$ and Mother has $2n$ socks. After giving one sock to his wife, Mother has $2n + 1$ and Father has $3n - 1$. Mother's $2n + 1$ is divisible by 5 due to her ability to share equally among five people, and since the total number of socks in the five boxes $5n$ is divisible by 5, then Father's $3n - 1$ must also be divisible by 5. Father's $3n - 1$ also has a factor of 8 due to his ability to share equally amongst eight people (including himself). The only possible values of $3n - 1$ less than 100 are 40 and 80. Only the second gives a whole number, $n = 27$.

4.22 Fruit and nut

The pairs involved in the interchanges are 8–2, 3–9, and 1–7. There are only two sets of number combinations of the oranges in the boxes that satisfy the given conditions. One is 384, 519, 762, and the other is 942, 186, and 537. The first set is the given one and so this must be transformed into the second set. There are two ways of doing this so that the non-participating compartments total 15: 8–2, 3–9, and 1–7; with 4, 5, 6 unmoved; or 3–1, 4–6, and 9–7 with 8, 5, 2 unmoved.

  The first case has no pair totalling more than 12 (valid) whereas the second case does (invalid).

4.23 Lettuce play

The two-digit pass number is 92. Being mutually contradictory, only one of statements (1), (3), and (7) can be true. So there are three true statements in the four remaining ones: (2), (4), (5), and (6). Using the method of exhaustion, there are 12 possible two-digit pairs to check (regardless of order) for these three statements. For each possibility we count up how many of the remaining four statements it satisfies. We discover that exactly three statements are true only for the pair 2 and 9. That gives the four true statements required out of the seven, namely, (2), (5), (6), and (7). With the higher digit preceding the lower one, the solution follows.

4.24 Going bananas

A dark grey (D) monkey is at A and a white (W) at B. Let a light grey be L. Adding moments (distance times weight), on the bottom left beam, $2 \times L = 1 \times W$. So $W = 2L$. The beam weight is 4L. On the bottom right beam, $1 \times D = 1 \times W + 2 \times W$. So $D = 3W = 6L$. The beam weight is 16L. The known moment on the top left beam is $4 \times L + 3 \times 2L + 2 \times 4L + 1 \times 6L = 24L$. The known moment on the top right beam is $1 \times 2L + 1 \times 16L + 2 \times 6L = 30L$. So with D at A we add $2 \times 6L = 12L$ to the left, and with W at B we add $3 \times 2L = 6L$ to the right, making 36L each side.

4.25 No pain no gain

Antwit is on rung 8, Babble on 7, Crumble on 1, Dibdib on 6, and Earwig on 9. From (1), (3), and (4) we have $A + 1 = E$, $D + 2 = A$, and $B = D + 1$ or $B = D - 1$. This allows (A, B, C, D, E) = (3, 2, C, 1, 4), (4, 1/3, C, 2, 5), (5, 2/4, C, 3, 6), (6, 3/5, C, 4, 7), (7, 4/6, C, 5, 8) or (8, 5/7, C, 6, 9). Using (2), we can have (4, 3, 1, 2, 5), (5, 4, 1, 3, 6), (6, 5, 1, 4, 7), (7, 6, 1, 5, 8) or (8, 7, 1, 6, 9). Finally, (5) reduces the possibilities to (8, 7, 1, 6, 9).

4.26 Milky weigh

The values are B = 4, E = 5, F = 0, and G = 1. This is an arithmetic moments problem that could be done by trial and error. However, it can also be done by algebra as follows. Let the weight of the coin on the left be x and the two coin weights on the right be y and z. For the coin on the left at A, three possible equations can be constructed (1) $3x = y + z$ (both at E), (2) $3x = 2y + 2z$ (both at F), or (3) $3x = 3y + 3z$ (both at G). For the coin on the left at B, we can have (4) $2x = y + 2z$ (at E and F), (5) $2x = y + 3z$ (at E and G), or (6) $2x = 2y + 3z$ (at F and G). If we examine the nine pairs formed from each of (1)–(3) with each of (4)–(6), only (2) with (5) works, so that $x = 4z$ and $y = 5z$. For the other eight combinations, either two values are identical, or a negative value occurs. Only $z = 1$ allows digital values for x and y less than 10, so $x = 4$, $y = 5$, and $z = 1$ and since they arise from (5), then y and z are at E and G.

4.27 Bird brain

A is 2, B is 6, C is 9, D is 3, E is 8, and F is 4. We note that $C = 3D$ so (C, D) can be (6, 2) or (9, 3). Also $E = 2F$ so (E, F) can be (4, 2), (6, 3) or (8, 4). So, (C, D, E, F) can be (6, 2, 8, 4), (9, 3, 4, 2) or (9, 3, 8, 4). On the top perch, the weight times distance (moment from the middle) of the bottom left perch (L) and the bottom right perch (R) produce (L, R) possibilities (24, 24), (36, 12) or (36, 24), respectively. Since A and B must be different weights the first case is eliminated, as is the second since no A and B allows balance. (In the second case (36, 12), the highest weights left for B are 7 and 8 and 3 × 7 = 21 and 3 × 8 = 24 allowing a maximum of 36 on the right with A still to be assigned.) The third case only allows A as 2 and B as 6, with a total moment of 42 each side.

4.28 Three shades of grey

The ratio is 1 : 4 : 7.
For brevity let light grey be L, white be W, and dark grey be D.
Initially number of units are L=1, W=2, D=3.
(1) One unit W to L leaving L=2, W=1, D=3
Total number of units of L, W, D in each glass.
L: 1L + 1W
W: 1W
D: 3D
(2) One unit L to D leaving L=1, W=1, D=4
L: (1/2)L + (1/2)W
W: 1W
D: (1/2)L + (1/2)W + 3D
(3) Two units D to W leaving L=1, W=3, D=2
L: (1/2)L + (1/2)W
W: (1/4)L + (5/4)W + (3/2)D
D: (1/4)L + (1/4)W + (3/2)D
(4) One unit W to L leaving L=2, W=2, D=2
L: (7/12)L + (11/12)W + (1/2)D
W: (1/6)L + (5/6)W + 1D
D: (1/4)L + (1/4)W + (3/2)D
(5) One unit L to D leaving L=1, W=2, D=3
L: (7/24)L + (11/24)W + (1/4)D
W: (1/6)L + (5/6)W + 1D
D: (13/24)L + (17/24)W + (7/4)D
So we examine the D fractions in each glass to find L: 1/4, W: 4/4, D: 7/4.

4.29 The odd getting even

The total is 84 odd number arrangements. Let a prime digit be called P = (2, 3, 5) and a non-prime digit be called N = (1, 4, 6). The possible arrangements to avoid adjacent primes are then (i) NPNPNP, (ii) PNNPNP, (iii) PNPNNP,

(iv) PNPNPN. In each of the three cases (i)–(iii), there are six ways the non-primes can be arranged amongst the three positions with N. With only two odd numbers possible for the last P place, there are four ways the primes can be distributed amongst the three positions P. This gives $3 \times 4 \times 6 = 72$ possibilities. For case (iv), there are six ways the primes can be arranged in the three positions P. Only one odd number is available for the last N place with 2 arrangements in the other positions N. This gives $1 \times 2 \times 6 = 12$. So Sam can choose from a total of $72+12 = 84$ possible odd numbers.

Reference

[1] Clarke, Barry R. *Brain Busters: Mind Stretching Puzzles in Math and Logic.* New York: Dover Publications Inc., 2003, p.19.

5 Algebra

5.1 Paying in nuts

In Weeble Wood, where the yellow squirrels live, they pay their debts with nuts. A hazelnut is worth 7 units, an acorn is worth 4 units, and a monkey nut is worth 3 units. Flossy the squirrel recently bought a portable DVD player for 63 units using 15 nuts. However, she couldn't recall how many of each type she used to pay for it. Each of the three numbers of coins was a single digit from 1 to 9 inclusive, no two being the same.

How many of each nut did Flossy use?

DOI: 10.1201/9781003358275-5

5.2 A calculated correction

Miss Flogginum, the mathematics teacher, is strict about talking in class and has marched three of the culprits to the front. Ever since corporal punishment was abolished, she has taken pleasure in administering her own form of justice. From left to right, Antwit, Babble, and Crumble are made to stand facing away from the class in a straight line, and each has an unseen digit from 1 to 9 pinned to their back, no two of the three digits being repeated. No pupil can see their own digit and, in turn, each is required to make a correct statement about the two-digit number that they can see on the backs of the other two.

"I see a number divisible by nine," said Antwit.

"I see a prime number," reported Babble.

Crumble was more thoughtful and impressed the class with the following observation. "Considering the two-digit number that I can see, subtracting from it the one formed by reversing its digits results in a positive number that is three times the sum of the two digits."

What digit did each pupil have?

5.3 Prime of life

Professor Neuron was contemplating his life in relation to the ages of three of his family members Agatha, Babble, and Clump.

When Neuron gained his doctorate 28 years ago, Babble was twice the age that Agatha was four years ago when Neuron became a professor. Even more interesting, thought Neuron, is the fact that when he retires 11 years from now, Agatha will be three times the age that Clump will be nine years hence, when the professor gets his free bus pass.

A total of five digits is required to construct the ages in years of Agatha, Babble, and Clump, and no two digits are repeated. Each age is presently a non–zero whole number and Agatha's current age is a prime number.

What are the present ages of Agatha, Babble, and Clump?

5.4 Digit ale dilemma

Down at the Pig and Bucket, Nibble was trying to interest his friend Muddle in a puzzle.

Said Nibble,

> "The number of bubbles in my beer, let's call it A, is a three–digit number constructed from the digits 1–9, no two being equal. If two of the three digits are juxtaposed to make a second number B, and B is subtracted from A, the result is a positive two–digit number C."

"Fascinating," said Muddle, "but it's your round."

Nibble was unperturbed. "Both of the digits found in C appear in A, and the sum of the digits in C is the remaining digit in A."

"Oh, and I'll have a packet of pork scratchings too," said Muddle.

What was the number A that Nibble had in mind?

5.5 Tridems and Quadems

At the Intergalactic Space Party only members from two tribes of aliens are present. These are the Tridems with three arms each, and the Quadems with four arms each. The ratio of Tridems to Quadems is determined by an agreement that there are to be equal total numbers of arms for each tribe at the proceedings.

It is a rule of the party that each guest must greet every other guest whether from the same tribe or not. During a greeting, each hand of a guest shakes every hand of another guest, each arm having only one hand.

What is the minimum number of handshakes that can occur at the party?

5.6 Fraudulent fruit

"Two hundred apples!" cried Shifty Sid, as he stacked up the ten boxes as shown. However, after he had gone, the market trader found that they actually totalled a square number less than 200.

As the boxes were examined in the order A to D, adjacent numbers in those four boxes were found to increase by the same difference which was greater than one. Also, apart from the bottom row, the number of apples in a box was the total of the two boxes immediately below it.

How many apples were in box H?

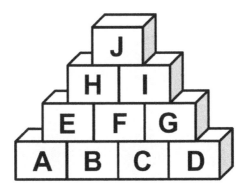

Figure 5.1   Boxes of apples.

5.7 Train of thought

On a normal day, the Algebraic Express (A) takes a time T (minutes), which is more than a quarter of an hour, to travel the distance d (km) at constant speed v (km/min). Unfortunately, the Bolting Bracket (B) is now approaching along the same line at a speed mv (km/min). It takes t (mins) for A to travel the distance x (km), and the same time for B to cover the distance y (km), after which time they collide. Now $(x + y) = d$, $T = nt$, and $v = a/b$, so that a, b, d, m, n, t, x, y are all whole numbers from 2 to 10 inclusive, with no two numbers being the same.

What is the distance x (km) and the time t (mins) for the collision?

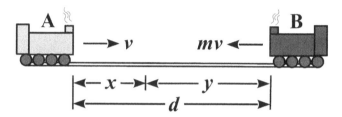

Figure 5.2   The colliding trains.

5.8 Duels for jewels

In Whimsy Wood, the creatures that live under leaves were holding their annual sports day event. Ollie the owl was ready to use his considerable hoot as a starting pistol, while Mollie the magpie had volunteered to present the jewellery prizes.

The race track was x (cm) long and was marked out in a straight line on the ground. Each race had two competitors, and despite Kevin the caterpillar, who began on the starting line, giving Sid the snail y (cm) start they still succeeded in crossing the finishing line simultaneously. In another race, when Ernie the earthworm began on the starting line, he tied with Bill the beetle who stood $7y$ (cm) behind the line. It turns out that Bill was as fast as Kevin, and Ernie was as fast as Sid.

Now x is a two-digit whole number and y is a square number.

What is the length x (cm) of the race track?

5.9 Frame game

Professor Neuron has ordered a rectangular wooden picture frame consisting of four wooden pieces (see Figure 5.3, not to scale). The dimensions of the large rectangle are $A = (a + x)$ cm and $B = (b + x)$ cm while those of the small one are $C = (b − x)$ cm and $D = (a − x)$ cm? Now, the professor has requested that the area of the wooden frame (grey region) should be precisely 60 cm^2. Also a, b, and x cm must be digits from 1 to 9 so that no two are the same. Evidently length B is greater than A, and no wooden piece is wider than 3cm. Here, the width of a piece means the horizontal or vertical distance across the wood from the inside to the outside of the frame.

Can you find the values of a, b, and x?

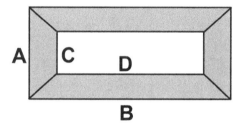

Figure 5.3   The wooden picture frame.

5.10 Try angling

Professor Neuron thought his nephew Trifle was spending far too much time on his smartphone, so he decided to give him a sum of money towards a new fishing rod. Luckily, Neuron had just discovered an interesting right-angled triangle from which he could calculate the amount he wished to give.

The length of the hypotenuse in Neuron's triangle he called A. When the lengths of the other two sides B and C were multiplied together, the result was the square of A divided by one less than the square. However, when B was divided by C, the answer was one greater than A divided by one less than A.

Neuron wanted to give his nephew a whole number of pounds, and he realised that he could make this number from his triangle by squaring the number A, subtracting 2, then squaring the result.

How much did Professor Neuron give his nephew towards a new fishing rod?

5.11 Digital differences

When families gather together, some like to argue about family indiscretions, others like to contest each other's politics, but the differences that divide opinion in the Digital family are usually the product of a comparison of some of their ages.

Said Adder to Bracket:

"When you were running around as a small child 12 years ago, I was three times the age you will be three years before you'll be twice your age now."

Said Cary to Bracket:

"Well, taking our ages now, the sum of my age and your age is Adder's age."

The present ages of Adder, Bracket, and Cary are each two-digit non-prime whole numbers?

How old was Cary when Bracket was born?

5.12 Frozen in time

It is a little known fact that over a century ago, Ernie Chuckletin, the polar explorer, set off on a heroic solo expedition across the arctic ice taking only a small supply of food. His intention was to leave basecamp, hike as far as possible along his chosen route, then return the same way.

The plan was to take a rest period after each hiking stage, a hike lasting for exactly two hours, and a rest period for exactly 15 minutes. A constant speed was maintained at each stage but, to allow for fatigue, he reduced his speed by one twentieth of his previous one, so that his first return hike was slower than his last outward one.

Ernie's team at basecamp had calculated that under these conditions, there was a maximum number of outward hikes above which there would be no prospect of his return. Fortunately, Arnie took the greatest time possible for his outward journey, and returned to base camp freezing cold but safe and well!

How much time did Ernie take after setting off before turning back?

5.13 Gift rap

At the Packet School for gifted children, Stamp and De Liva were wrapped up in a problem. There were three school store rooms A, B, and C. Room A contained a number a of empty boxes, B contained b, and C contained c. The two children were discussing ways of constructing the number c.

Said Stamp, "if we take four times a and add it to three times b we get a number X."

"Yes," said De Liva, "and if we take eleven times b and subtract it from seven times a, we get another number Y."

"Interesting," said Stamp, "because if we now multiply X and Y together we get the number c."

The numbers a and b were each a single digit from 1 to 9 inclusive, and the value of c was a two-digit prime number.

How many boxes are in Room C?

5.14 Loco-motive

Due to a signals failure, the unmanned Bolting Bracket (B) is hurtling towards the passenger train, the Chortling Chuffer (C), on the same railway line. The engine driver on the Algebraic Express (A) has decided to pursue train B along the line, board it, and apply the brakes before it strikes C. This should give train C a greater chance of stopping in time.

As we join the drama, engine A is moving at speed a km/min, B at b km/min, and C at c km/min. Also, A is e km from B, which is f km from C. If all remain constant during the chase, A will catch B in g min, after which time the latter engine will have travelled a distance d km.

The numbers a, b, c, d, e, f, g are each a single digit 1–9, no two digits being the same. Also distance d is less than e, and the lengths of the engines are assumed to be negligible compared with the distances travelled.

What are the speeds a, b, c, of the three engines?

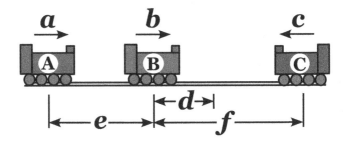

Figure 5.4   Trains A, B, and C.

5.15 Tea 'n age

Antwit, Babble, and Crumble are three teenagers such that no two of their whole-number ages are the same. Often, when they feel like they're negotiating an existential crisis, they sit down and discuss their ages over a pot of camomile tea. Being students of mathematics, their conversation on such matters usually takes the form of a calculation.

Said Crumble to Babble, "five years ago, I was half the age you will be in four years time."

Said Antwit to Crumble, "in the year to which you refer, I was twice the age you were three years before that."

It is difficult to see how they felt existentially enlightened as a result, but it cannot be in doubt that mathematicians have a curious way of amusing themselves.

What are the present ages of Antwit, Babble, and Crumble?

5.16 Sense of direction

Professor Neuron was apt to tease his nephew Trifle, not as an attempt to humiliate, but rather as a way of encouraging his mind to engage with the mysteries of the universe. So as they shared a pot of camomile tea in the professor's study, Trifle was expecting his uncle to deliver yet another mind-stretching conundrum.

"Let us consider forwards," said Neuron.

"You mean the four words you've just spoken?" said Trifle, delighted with his own pun.

"And let us consider backwards," said Neuron, reaching into his pocket.

"In what sense might there be no difference between the difference between forwards and backwards, and either one of forwards and backwards?"

Although Trifle was not backwards at going forwards he was now thoroughly confused. With that, Neuron produced a piece of paper and carefully unfolded it.

Let me explain. On this piece of paper, I've written a three-digit number using the digits 1–9. No two of these digits are the same. The digits must be reversed and the result subtracted from the original number. This creates a positive number containing the same three digits that appear in the original number.

What number was written on the piece of paper?

5.17 Strawberries and crime

Professor Neuron and his nephew Trifle have been out in the garden collecting berries. The professor has picked a number of strawberries while young Trifle has brought back a smaller number of blackberries. The sum of their two numbers, for those who take delight in such matters, is seven times their prime-number difference.

Now, while the professor isn't looking, Trifle creeps into his study and steals a number of his strawberries. However, the professor is very perceptive, tiptoes into Trifle's room, and takes some of his blackberries. Neuron now decides to confront Trifle in the form of a poser.

"If I take twice the number of berries of any type I now have, and add three times the number that is 26 less than the number you have just taken, then that would amount to 2 less than the number of any type that you now have. So, how many would that be?"

"Well, I'm not sure," said Trifle, realising the game was up. However, he soon saw the amusing side of it all when he realised that his uncle had taken a single-digit number of his blackberries by way of revenge.

What is the sum of the number of berries of any type that Neuron and Trifle have collected between them?

5.18 Panic in the Pentagon

Bricklayers Luis, Juan, Francesco, Pedro, and Jorge stand at the vertices of a pentagon, each with a pile of 10 bricks. To build strength, they train by throwing bricks for each other to catch in the order given below. Of course, dropping one could result in injury so all five workers are fully focused.

(1) Luis throws *a* bricks to Juan and Jorge throws the same number to Francesco.
(2) Juan throws *b* to Francesco and Luis throws the same number to Pedro.
(3) Francesco throws *c* to Pedro and Juan throws that number to Jorge.
(4) Francesco throws *d* to Luis and Pedro throws the same to Jorge.
(5) Pedro throws *e* to Juan and Jorge throws that number to Luis.

This leaves Luis with 11 bricks, Juan with 18, Francesco with 11, Pedro with 6, and Jorge with 4. Figure 5.5 shows the final piles of bricks marked by the first three letters of the thrower's name together with the final total in each pile and the direction of the throws. The digits *a*, *b*, *c*, *d*, *e* are from 1 to 9 inclusive, and at no time did anyone have no bricks.

What are the values of a, b, c, d, and e?

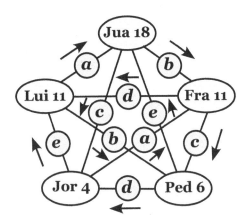

Figure 5.5   The directions of brick-throwing and the final brick numbers.

Solutions

5.1 Paying in nuts

He used 4 hazelnuts, 2 acorns, and 9 monkey nuts. Let there be h hazelnuts, a acorns, and m monkey nuts. So $7h + 4a + 3m = 63$ and $h + a + m = 15$. Multiplying the second equation by 7 and subtracting the first leads to $a = (42 - 4m)/3$. Running up the m we find that single digits for all three of h, a, m occur for $m = 6$ and $m = 9$, but in the first case $a = 6$ (invalid duplication). This leaves the first case, $m = 9$, $a = 2$, $h = 4$.

5.2 A calculated correction

Antwit had 4, Babble had 2, and Crumble had 7. Let the three digits on the backs of Antwit, Babble, and Crumble be A, B, and C, respectively. Crumble's statement, which is about A and B, gives $10A + B - (10B + A) = 3(A + B)$ so that after expansion and simplifying we have $A = 2B$. This allows 21_, 42_, 63_, or 84_. Combined with Babble's statement, which is about A and C, we can have 213, 219, 421, 423, 427, 631, 637, 843, or 849. Antwit's statement, which is about B and C, only allows 427.

5.3 Prime of life

Agatha is 31, Babble is 82, and Clump is 5 years old. Let their present ages be A, B, and C, respectively. Then $B - 28 = 2(A - 4)$ and $A + 11 = 3(C + 9)$ which leads to $B = 6C + 52$. If there is to be a total of five digits in the three ages, then C is a whole number with $1 \leq C \leq 7$, B rises in increments of 6 where $58 \leq B \leq 94$, and A rises in increments of 3 where $19 \leq A \leq 37$. A prime number occurs for A when $C = 1$, $B = 58$, $A = 19$; $C = 5$, $B = 82$, $A = 31$; and $C = 7$, $B = 94$, $A = 37$. Only for $C = 5$ are no two digits of the five digits the same.

5.4 Digit ale dilemma

The number is 594. Let the digits be a, b, and c. The number A can then be written as $100a + 10b + c$. The possible juxtapositions for B are a with b, a with c, or b with c. The results of $C = B - A$ are then $90(a - b)$, $99(a - c)$, or $9(b - c)$, respectively. A two-digit number C containing two different digits 1–9 is only possible using the last choice because for the first two, the only possibilities are $a - b = 1$ ($C = 90$) and $a - c = 1$ ($C = 99$) which lead to an invalid C. For the third choice, we tabulate the two-digit values of C, and the possible abc as follows.

Only for 594 is the third digit (9) the sum of the other two in C (4 and 5).

Table 5.1 The differences $(b - c)$ determining the two-digit values of C and the possible abc

$b - c$	2	3	4	5	6	7	8
$C = 9(b - c)$	18	27	36	45	54	63	72
Possible abc	186/831	274/752	362/673	594	none	none	none

5.5 Tridems and Quadems

The minimum number is 246. Let there be a Tridems with 3 arms each and b Quadems with 4 arms each. For equal numbers of arms $3a = 4b$ so the smallest numbers of guests must be $a = 4$ and $b = 3$. When two Tridems greet, there are $3 \times 3 = 9$ handshakes, two Quadems $4 \times 4 = 16$, and when Tridems greet Quadems $3 \times 4 = 12$. Each of the a Tridem interacts with $(a - 1)$ other Tridems, and there are 9 handshakes between two Tridems, but we must divide by 2 to avoid double counting. So there are $a(a - 1)/2$ greetings between Tridems, $b(b - 1)/2$ between Quadems, and ab between both tribes. The total number N of handshakes is then $N = 9a(a - 1)/2 + 12ab + 8b(b - 1)$. The solution arises from $a = 4$ and $b = 3$.

5.6 Fraudulent fruit

There were 20 apples in box H.

Let box A have a apples, and let d be the common difference in the bottom row, which is greater than one. Then B has $(a + d)$, C has $(a + 2d)$, and

Table 5.2 Possible totals in the 10 boxes
depending on the numbers in A and B

a + d	(a, d)	N = 26a + 39d
2	(0, 2)	78
3	(0, 3)	117
	(1, 2)	104
4	(0, 4)	156
	(1, 3)	143
	(2, 2)	130
5	(0, 5)	195
	(1, 4)	182
	(2, 3)	169
	(3, 2)	156
6	(3, 3)	195
	(4, 2)	182

D has $(a + 3d)$. We then find that the total N in the 10 boxes is $N = 26a + 39d$. Pairs of values (a, d) can now be systematically checked for N being a square number less than 200 by first selecting $a + d = 2$ with a possible pair $(a, d) = (0, 2)$, then $a + d = 3$ with the possibilities $(0, 3)$ and $(1, 2)$, and so on. Table 5.2 shows the possibilities with the totals for N less than 200. For $a + d = 5$ and $(a, d) = (2, 3)$ we find $N = 26 \times 2 + 39 \times 3 = 169$ (13 squared). Since box H contains $(4a + 4d)$ then this amounts to $4 \times 2 + 4 \times 3 = 20$ apples.

5.7 Train of thought

The collision occurs at $x = 2$ km at time $t = 6$ mins. This could be a programmable problem but with some concentration a calculation on paper is possible. Since A and B collide after travelling t mins then $t = x/v = (d - x)/(mv)$. So rearranging gives $(m + 1)x = d$. The possible combinations of values for (m, x, d, y) are (i) $(2, 3, 6, 9)$, (ii) $(3, 2, 8, 6)$, and (iii) $(4, 2, 10, 8)$. Since $v = a/b$ (given) and $t = x/v$ then $t = bx/a$. For each case, we take the value of x and look for solutions for (a, b, x, t) taking care not to duplicate values. For (i) we have $t = 3b/a$ but no combination of (a, b, x, t) is possible. For (ii), $t = 2b/a$ and we can have $(4, 10, 2, 5)$ and $(5, 10, 2, 4)$. For (iii), $t = 2b/a$ and we can have $(3, 9, 2, 6)$, $(6, 3, 2, 1)$, and $(6, 9,$

2, 3). Considering that $T = d/v = bd/a = nt$ we arrive at $n = bd/(at)$. Since n is a positive integer such that $2 \le n \le 10$ then we find that the possible sets (m, x, d, y, a, b, t, n) are (4, 2, 10, 8, 3, 9, 6, 5), (4, 2, 10, 8, 6, 3, 1, 5), and (4, 2, 10, 8, 6, 9, 3, 5). The final check is for $T > 15$ (given). Since $T = nt$ then only the first case applies.

5.8 Duel for jewels

The length of the race track is $x = 42$ (cm). Let v_b, v_c, v_e, and v_s be the speeds of the beetle, caterpillar, earthworm, and snail, respectively. The first race with equal times gives the relation $v_s/v_c = (x - y)/x$. The second with equal times produces $v_b/v_e = (x + 7y)/x$. Since $v_b = v_c$ and $v_e = v_s$, the first ratio multiplied by the second must give $(x - y)(x + 7y)/x^2 = 1$ and leads to the result that either $y = 0$ or $x = 7y/6$. Given that y is a square number, the first solution is ruled out, and if x is a whole number then y must be divisible by 6. To ensure that x is two digits we can only have $y = 36$ so that $x = 42$.

5.9 Frame game

The values are $a = 7$, $b = 8$, and $x = 2$. The area of wood is $(a + x)(b + x) - (a - x)(b - x) = 60$. This simplifies to $x(a + b) = 30$. We now test factors x and $(a + b)$ of 30. Table 5.3 gives the possibilities, keeping in mind that $B > A$, no two of a, b, x are identical, $a > x$, and $b > x$.

The frame width requires that $(B - D)/2 \le 3$ and $(A - C)/2 \le 3$. Only the first possibility is allowed.

Table 5.3 Frame dimensions depending on the factors x and $(a + b)$

x	a + b	a	b	A	B	C	D
2	15	7	8	9	10	6	5
2	15	6	9	8	11	7	4
3	10	4	6	7	9	3	1

5.10 Try angling

Neuron gave his nephew £5 towards a new fishing rod, which makes one wonder which of them is the more disconnected from reality! From the given information, $BC = A^2/(A^2 - 1)$ and $B/C = (A + 1)/(A - 1)$. Substituting the second of these into the first reveals that $B = A/(A - 1)$ and $C = A/(A + 1)$. Being a right-angled triangle we have the Pythagorean condition that $A^2 = B^2 + C^2$ which leads to $A^4 - 4A^2 - 1 = 0$. The desired amount for Neuron's nephew is $(A^2 - 2)^2$ which arises from adding 5 to both sides of the quartic equation to give $(A^2 - 2)^2 = 5$.

5.11 Digital differences

Cary was 63 years old at Bracket's birth. Let Adder's present age be A, Bracket's be B, and Cary's C. Then Cary's pronouncement means that $A = B + C$. Adder's assertion leads to $A - 12 = 3(2B - 3)$ and so $A = 6B + 3$. We now list the possibilities for the two-digit numbers (A, B, C) starting with the first two-digit whole number for B's present age greater than 12 (which is 13) because he must have been at least one year old 12 years ago if he was then "running around" (toddlers typically start running at 18 months). So we have (81, 13, 68), (87, 14, 73), (93, 15, 78), or (99, 16, 83). Since 13 in the first, 73 in the second, and 83 in the last are prime numbers (invalid), then we can only have (93, 15, 78). So at Bracket's birth 15 years ago, Cary was $78 - 15 = 63$.

5.12 Frozen in time

Ernie set off back after 29 hours and 15 minutes (29.25 hours). For the distances travelled, the problem involves a geometric series, with first term $a = vt$ (velocity v multiplied by time $t = 2$ hours) and common ratio $r = 19/20$. With n as the number of outward stages, the sum of the return distances $vtr^n/(1 - r)$, which is the sum to infinity minus the sum to n terms, must be greater than or equal to the sum of the outward distances $vt(1 - r^n)/(1 - r)$. After cancellations, this leads to $(19/20)^n \geq 1/2$. Use of a calculator with logarithms now gives $n \leq 13.513$ so that $n = 13$. His 13 outward hikes take $2 \times 13 = 26$ hours and his 13 rest periods take $13 \times 0.25 = 3.25$ hours which when totalled give the solution.

5.13 Gift rap

There are 47 boxes in Room C. We can form the relation $(4a + 3b)(7a - 11b) = c$. Since c is a prime number, one of the numbers in parentheses must be equal to 1. However, a and b are each at least 1 so it must be $(7a - 11b) = 1$ and rearrangement gives $b = (7a - 1)/11$. Substituting this into $(4a + 3b)$

in the multiplication gives $c = (65a − 3)/11$. Running up the possible values of a starting at $a = 1$ reveals that only when $a = 8$ does a two-digit prime number occur which is $c = 47$. Since $(7a − 11b) = 1$ then $b = 5$. There is a fascinating relationship between a, b, and c. In $(4a + 3b)(7a − 11b) = c$, we note that the digits in $c = 47$ are the coefficients of a, and the difference and sum of the digits in c are the coefficients of b.

5.14 Loco-motive

We have $a = 7$, $b = 3$, $c = 1$, $d = 6$, $e = 8$, $f = 9$, $g = 2$. For A to collide with B, they must have equal times of travel to the rendezvous point a distance d from B's initial position, so

$$(e + d)/a = d/b = g \tag{5.1}$$

In Table 5.4 we list the possible single-digit times g mins that are possible from b and d, keeping in mind that no digit is repeated.

So, the possible solutions are given in Table 5.5, where g_{A-B} is the time for A to reach B.

The time that C takes to reach the stopping point of B is

$$g_{B-C} = \frac{f - d}{c} \tag{5.2}$$

where g_{B-C} is the time for C to reach B. In Table 5.5, d can be 6 or 8, so in Table 5.6 we list the possible times g_{B-C} times which must be greater than g_{A-B}.

So, the possible solutions from Table 5.6 are given in Table 5.7.

This means that due to the value of d, we are restricted to combining Table 5.7 with either the first or second row in Table 5.5, to form Table 5.8.

Table 5.4 Possible values in of g

g	d/b	$g_1 = (e + d)/a$	$g_2 = (e + d)/a$
2	6/3	$(e + 6)/a = (4 + 6)/5$	$(e + 6)/a = (8 + 6)/7$
2	8/4	$(e + 8)/a = (6 + 8)/7$	
3	6/2	$(e + 6)/a = (9 + 6)/5$	

Table 5.5 Possible combinations of a, b, d, e, g

A	b	c	d	e	f	$g_{A\text{-}B}$
5	3		6	4		2
7	3		6	8		2
7	4		8	6		2
5	2		6	9		3

Table 5.6 Possible values in (5.2)

D	$g_{B\text{-}C} = (f - d)/c$	Possible f values	$g_{B\text{-}C} > g_{A\text{-}B}$
6	$(f - 6)/c$	7, 8, 9	$(9 - 6)/1 = 3 > 2$
8	$(f - 8)/c$	9	none

Table 5.7 Possible combinations of c, d, f, $g_{B\text{-}C}$

a	b	c	d	e	f	$g_{B\text{-}C}$
		1	6		9	3

Table 5.8 Possible combinations of a–g

a	b	c	d	e	f	$g_{A\text{-}B}$	$g_{B\text{-}C}$
5	3	1	6	4	9	2	3
7	3	1	6	8	9	2	3

With the condition that e is greater than d, only the second row applies. We note that $g_{B\text{-}C}$ is not one of the digits of the set in which no digit is repeated.

5.15 Tea 'n age

Antwit is 19, Babble is 16, and Crumble is 15. Taking the initial of each name for the age, we have $C - 5 = (B + 4)/2$ and $A - 5 = 2(C - 8)$. This gives $2C - 16 = A - 5 = B - 2$. If we now run up the whole number ages 13 to 19 for A, we discover that C is only a teenager for A = 15, 17 and 19. However, B is only 12 for the first. The second gives equal ages for B and C. So the only possible triplet (A, B, C) with A, B, C as teenagers with no two the same is (19, 16, 15).

5.16 Sense of direction

The solution is 954. Let the original number be written as $100A + 10B + C$. Then the result of reversal to $100C + 10B + A$ and subtraction is $X = 99(A - C)$. Let us go through the three-digit results X for the possible differences $A - C = 2, 3, ..., 9$. These are $X = 198, 297, 396, 495, 594, 693, 792, 891$, respectively, which all contain the digit 9. Either B is 9 or one of A and C is 9. Let us suppose $B = 9$ in the original number so that the other two digits are A and C, order unknown. Then in no case is the difference between the larger and smaller of the end digits in X, which must be A and C with the order unknown, equal to $A - C$. So one of A and C is 9. It must $A = 9$ because the result of the subtraction $A - C$ is a positive X. So we now list the C in each case for $9 - C = 2, 3, ..., 9$ keeping in view the corresponding X in each case. The only occasion that C appears in its corresponding X is for $X = 495$ with $C = 4$, so we must have $B = 5$.

5.17 Strawberries and crime

The total number of berries is 77. Let Neuron's total be N, Trifle's T, and their difference D. Then $N - T = D$, $N + T = 7D$, so $N = 4D$ with $T = 3D$. Let Neuron lose n strawberries and Trifle t blackberries. Then $2(N - n + t) + 3(n - 26) = T - t + n - 2$. So $5D + 3t = 76$. The number D is defined as prime so testing 2, 3, 5, 7, 11, 13, the only single-digit whole number t is $t = 7$ when $D = 11$. So $N + T = 7D = 77$.

5.18 Panic in the pentagon

We have $a = 7$, $b = 1$, $c = 4$, $d = 3$, and $e = 6$.
   For Luis we have

$$10 + d + e - b - a = 11 \tag{5.3}$$

for Juan

$$10 + a + e - b - c = 18 \tag{5.4}$$

for Francesco

$$10 + a + b - c - d = 11 \tag{5.5}$$

for Pedro

$$10 + b + c - e - d = 6 \tag{5.6}$$

and for Jorge

$$10 + c + d - a - e = 4 \tag{5.7}$$

Adding (5.3) and (5.5) give $e - c = 2$; (5.3) and (5.6) give $a - c = 3$; (5.4) and (5.6) give $a - d = 4$; (5.4) and (5.7) give $d - b = 2$; and (5.5) and (5.7) $e - b = 5$. Adding the third and fourth of these gives $a - b = 6$. The only possibilities for (a, b, c, d, e) are $(7, 1, 4, 3, 6)$, $(8, 2, 5, 4, 7)$, and $(9, 3, 6, 5, 8)$. Only the first case is possible because, with the given order, Luis is left with zero or less bricks on the fourth throw with the other two cases.

Programmable puzzles

6.1 Sweet truth

Antwit, Babble, Crumble, and Dibdib each have a number of chocolates 1–9 inclusive, no two numbers being the same. The sum of Antwit and Babble's share divided by the sum of Babble and Crumble's give Dibdib's number. The sum of their four digits is a square number.

How many chocolates do Antwit, Babble, Crumble, and Dibdib have?

DOI: 10.1201/9781003358275-6

6.2 Neuron's number

During COVID-19 lock-down, Professor Neuron found a wonderful way to waste his time. He threw a regular six-sided dice three times until he had three different digits 1–6. Then he wrote them down as a three-digit number. On one occasion, neither of the two end digits was prime, the number was divisible by 3, and the third digit was less than the second one by more than 1.

What was the three-digit number?

Figure 6.1   A regular six-sided dice.

6.3 Squares and cubes

Professor Neuron is showing off again and has drawn two identical dice with the number of spots 1–6 on a face represented by a letter. The following relations apply.

(1) The sum of the spots on faces B and E is a square number x.
(2) The result of D subtracted from C is a square number y.
(3) The result of A subtracted form E is a square number z.
(4) No two of x, y, z are the same.
(5) The number of spots on F is greater than B.

Write a computer program to determine the following.

What are the numbers on the faces, F, A, C, E?

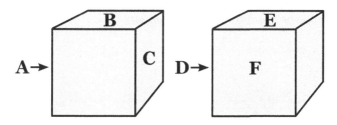

Figure 6.2   Labelling for the number of spots on the two dice.

6.4 The five digits

Professor Neuron has forgotten the five-digit code to the combination lock on his wallet. He knows that each digit is from 1 to 9 inclusive, and he also knows that no two digits are the same. Although this limits the number of permutations to 15,120, opening the lock by systematically and physically testing all possibilities is still a formidable task, even for someone with Neuron's numerical nous. Fortunately, he is able to recall the following facts about the correct code.

(1) Exactly two digits are square.
(2) The first and fifth differ by 2.
(3) The second and third total the fourth.
(4) The first and second differ by 1.
(5) The fifth digit is odd.
(6) The second and fourth differ by 1.

What is Professor Neuron's five-digit number?

6.5 Moggy maths

Each of the nine members of the Feline Fat cat family has a different digit value for its mass 1–9 kg. Three of them decide to sit on the see-saw (see Figure 6.3), one to a seat, so that it balances. The seats are set at one and two units distance from the middle. Their total mass is 15kg, and there is at least one prime number mass. Write a computer program to determine the following.

What is the sum of the masses of the two cats who sit on the same side?

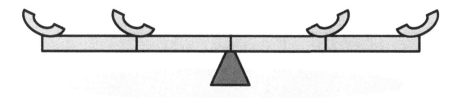

Figure 6.3   The seating arrangement on the see-saw.

6.6 Pizza problem

Muddle and Nibble were standing outside the baker's shop window comparing their finances.

Said Muddle, "I only keep 12-honk coins and I have one in my pocket."

Replied Nibble, "I only keep 45-honk coins and I have one in my pocket."

"That's a total of 57 honks," said Muddle, "not enough dough to buy that square piece of pizza in the window."

Said Nibble, who happened to be quick at arithmetic, "if we had a total of 47 coins and each had a square amount of honks, then together we would have the exact amount of honks required."

How many honks did the square of pizza in the window cost?

6.7 The benevolent banker

Bumbletown Bank has a number of gold coins in its vault, securely locked away behind heavy steel doors. For many years the bank has obtained a substantial profit from investing its clients' deposits. However, Mr Good, the manager, has now generously decided to give away the contents of the vault as a good-will gesture!

On Monday he distributes two-thirds of the gold coins to his customers, on Tuesday he hands out four-fifths of the remainder, and on Wednesday he donates six-sevenths of what is then left. The rest is spent on a lavish party for his loyal staff. The original number of coins is a square with six digits (0–9) so that no two digits are the same.

How many gold coins does the staff party cost?

6.8 Cake walk

Professor Neuron has been networking and has decided to hold a party for his new connections. At the bakery, he decides to buy a number of identical currant cakes, one for each of the attendees at that evening's party. The cakes are packed in either a white box or a grey box and the professor plans to carry the cakes home.

A white box contains A cake(s) and a grey box holds B cake(s). Neuron realises that if he carries C white boxes and D grey boxes he would have the same total number of cakes as he would have carrying E white boxes and F grey boxes. The values A, B, C, D, E, and F are each a digit from 1 to 6 inclusive, with no digit used more than once. Also D is greater than A which is greater than F.

How many will attend that evening's party?

Figure 6.4   The equation for the currant cakes.

6.9 Never odd or even

At Palindrome Palace in Scandinavia, the visitors must always enter the court of King Otto walking forwards, and leave it walking backwards. Also, since Otto takes pleasure in feeling superior, their first and last words must always be the palindromic cry, "Dammit, I'm mad!" For anyone who is neglectful, the penalty is eternal banishment; however, the transgressor is traditionally offered the opportunity of redemption if a digital palindromic puzzle can be solved.

Unfortunately, Barc Crab has just violated the king's protocol: his exit was made walking sideways instead of in reverse. So now a court official has presented him with the task of deciphering a palindromic number, that is, one that reads identically forwards and backwards.

The puzzle has seven digits, to be selected from 1 to 9 inclusive, which satisfy the following conditions.

(1) The 4th and 6th digits differ by 2.
(2) Either the 3rd or 7th digit is even but not both.
(3) There is only one square digit.
(4) Only two positions have a prime digit.
(5) The 5th and 6th digits differ by 3.
(6) No digit appears more than twice.

What is the seven-digit number?

6.10 Court out

The Great Mephisto has asked a volunteer to select four playing cards and lay them on the table, one face up, and three face down. Of these four cards, there are to be no court cards, only one of each suit, with no two values the same. His intention is to reveal the three face down cards. The trouble is, instead of using a marked deck, the magician has mistakenly used an ordinary one, and has no idea what the concealed cards are. However, the following facts are true about the four cards.

(1) The first card, which is not red, totals a square number with the last card.

(2) The sum of the club and diamond is at least 16.

(3) The club is to the right of the diamond.

(4) The sum of the first two cards is no more than 12.

(5) The four cards sum to a square number.

What is the third card from the left?

Figure 6.5   The arrangement of the four cards.

6.11 Food for thought

"We must keep ourselves in prime shape, eat proper food," said Professor Neuron to his nephew Trifle.

"Proper food?" said Trifle.

"Yes, proper food. And we'll take regular exercise. Every evening we'll walk a whole number of miles."

Despite Trifle's misgivings, they managed to keep to protein, fat, and vegetables, and reduce their carbohydrate intake. Also, on the first five days of COVID-19 lockdown they succeeded in walking a, b, c, d, and e miles respectively, where each of the five values is a digit 1–9. No two values are the same and the following relations hold true.

(1) c is odd
(2) a is one greater than e
(3) c and d total b
(4) exactly two digits are square
(5) b and c differ by one
(6) c is two less than e

Can you cut your carbohydrates and take sufficient exercise to discover the total number of miles that they walked over the first five days?!

6.12 Carts of coal

In Victorian London, there was a coal yard at the top of a steep hill which customers would visit with their horse and cart. The yard sold two different size bags of coal, one being a two-digit number ab (which represents $10a + b$) times the weight of the other. One day, Dirty Des turned up with his horse Neddy and asked for a number c of large bags and a number d of small bags to make a number e of hundredweight that his horse could pull. Shortly afterwards, Grimey Graham appeared with his horse Dobbin and asked for a number f of large bags and a number g of small bags to get the number h of hundredweight that Dobbin could pull.

Each of a, b, c, d, e, f, g, h was a single digit from 2 to 9 inclusive. No two digits being the same.

What were the weights of the large and small bags?

Solutions

6.1 Sweet truth

Antwit has 9, Babble has 1, Crumble has 4, and Dibdib has 2. Let their numbers be A, B, C, and D, respectively. The computer program can run in nested loops through the integer permutations $0 < A, B, C < 10$ and allow output only if A is not equal to B nor C, and B is not equal to C. Then set $D = (A + B)/(B + C)$ and allow output only if D is not equal to A, B, nor C. Also, output is given only if $A + B + C + D = 16$ or 25. This should give the solution.

6.2 Neuron's number

The number is 651. Let the first, second, and third digits be $p1$, $p2$, and $p3$, respectively. There can be three nested loops where each runs from 1 to 6. The conditions for output are as follows: (1) no two are the same, (2) $p1 + p2 + p3$ can be any multiple of 3 from 6 to 24, (3) $p2 - p3 > 1$, (4) create a counting variable X for the number of primes which is incremented by 1 if $p1$ is prime and again if $p3$ is prime, so that $X = 0$ for output.

6.3 Squares and cubes

$A = 2$, $B = 3$, $C = 5$, $D = 4$, $E = 6$, and $F = 5$. We must arrange to run through the integer permutations $0 < A, B, C, D, E, F < 7$ in nested loops. The following are the conditions for allowing output. We must have $F > B$. Let $x = B + E = 4$, 9, or 16. Let $y = C - D = 1$ or 4. Let $z = E - A = 1$ or 4. No two of x, y, and z are equal. Since A, B, and C are on the same dice, no two of them are equal. Again, D, E, and F are on the same dice so no two of them are equal. Opposite faces of a dice total 7, so $A + C = 7$ and we must not have $(A + B)$, $(B + C)$, $(D + E)$, $(D + F)$, nor $(E + F)$ totalling 7. This should uniquely determine the solution.

6.4 The five digits

The number is 78195. This could be completed by trial and error but a computer program written in Liberty BASIC can be found in the Appendix.

131

6.5 Moggy maths

Their total of 2 kg and 6 kg is 8 kg. The weights are 2 kg, 6 kg, and 7 kg. The moments calculation is $6 \times 2 + 2 \times 1 = 7 \times 2$. Four nested loops can be created with variables $p1$, $p2$, $p3$, and $p4$, each taking the integer values 0 to 9. Checks on these are as follows: (1) no two are equal, (2) exactly one digit must be zero corresponding to an empty seat, (3) their sum is 15, and (4) there are greater than zero prime values. The moments on one side can be written $L = p1*2 + p2$ and on the other side $R = p3 + p4*2$. There is a check for their equality. Only when all these conditions are satisfied is there any output allowed (all values are printed).

6.6 Pizza problem

The pizza cost 1224 honks. Muddle starts with $12 = 2 \times 2 \times 3$ honks and Nibble has $45 = 3 \times 3 \times 5$ honks. To make a square number of honks for each, Muddle needs $3m$ and Nibble $5n$ coins, where m and n are square numbers. In other words, there must be an even number of prime factors (2, 3, 5, etc.) for each person. Trying permutations of square numbers for $(m, n) = (1, 1), (1, 4), (4, 1), (4, 4), (1, 9), (9, 1), (4, 9), (9, 4)$ reveals that $m = 9$ produces 27 coins and 324 honks for Muddle, and $n = 4$ gives 20 coins and 900 honks for Nibble, a total of 47 coins. So the cost of the pizza is the total of these two amounts.

6.7 The benevolent banker

The party costs 8,505 gold coins. The six-digit number must be divisible by at least 3, 5, and 7 as prime factors. However, multiplication of the least square possible $N = 3^2 \times 5^2 \times 7^2$ by 11^2 produces a number with more than six digits so the square can only be constructed from the prime factors 2, 3, 5, and 7. For a six-digit number to occur, a test on N shows that while keeping the other powers as 2, the power of 3 cannot be greater than 6, and the powers of 5 and 7 cannot be greater than 4. With these restrictions, we look for a number of the form $2^a \times 3^b \times 5^c \times 7^d$ where a, b, c, d can each take even number values with zero allowed only for a. The only six-digit squares are 176400 ($a = 4$, $b = 2, c = 2, d = 2$), 275625 (0, 2, 4, 2), 396900 (2, 4, 2, 2), 540225 (0, 2, 2, 4), 705600 (6, 2, 2, 2), and 893025 (0, 6, 2, 2). Only the last has no two digits equal. So the solution arises from 893025 multiplied by $(1/3) \times (1/5) \times (1/7)$.

6.8 Cake walk

The total number of guests at the party is 27. The calculation can be written as $C \times A + D \times B = E \times A + F \times B$. This leads to $(C - E)/(F - D) = B/A$ which can be used to check alternatives. The only calculation that works is $1 \times 3 + 4 \times 6 = 5 \times 3 + 2 \times 6 = 27$.

6.9 Never odd or even?

The number is 8634368. Condition (6) means that the central digit is not repeated because if it is then it must appear three times. Every digit to the left of centre has a symmetrical duplicate to the right of centre. From (3), the fourth digit must be the only square number (1, 4, or 9). The problem can then be simplified by replacing 6th with 2nd in (1), 7th with 1st in (2), and 5th and 6th with 3rd and 2nd, respectively, in (5), and concentrating on the first four digits. From (3) and (4), it follows that only one of the first three digits can be a prime.

6.10 Court out

From left to right the cards are ace of spades, seven of hearts, nine of diamonds, and eight of clubs. From (1), since the first card is not red (diamond or heart) then it must be black (club or spade). Using (3), the first card cannot be a club, so the club is fourth, the spade is first, and the diamond is third.

The following conditions are now programmable. Let the card values from left to right be p_1, p_2, p_3, p_4. Each takes a whole number value from 1 to 10 inclusive. No two values are identical.

(i) $p_1 + p_4 = 4, 9,$ or 16
(ii) $p_3 + p_4 \geq 16$
(iii) $p_1 + p_2 \leq 12$
(iv) $p_1 + p_2 + p_3 + p_4 = 9, 16, 25,$ or 36

There are six possible solutions but only one has a seven in the second position: 1, 7, 9, 8.

6.11 Food for thought

They walked a total of 19 miles. The digits are $a = 6$, $b = 4$, $c = 3$, $d = 1$, and $e = 5$. As well as no two being equal, and each digit taking whole number values from 1 to 9, the programmable conditions are as follows.

(i) $c = 1, 3, 5, 7,$ *or* 9
(ii) $a = e + 1$
(iii) $b = c + d$
(iv) count the number of digits that are square and accept when greater than one
(v) $|b - c| = 1$
(vi) $c = e - 2$

6.12 Carts of coal

The small bag $x = 1/42$ cwt and the large bag $y = 16/21$ cwt. The problem is to find the different digits $a, b, c, \ldots, h$ in the eight-digit set 2, 3, 4, ..., 9 that allows a solution to the equations

$$y = abx$$
$$cx + dy = e \tag{6.1}$$
$$fx + gy = h$$

where ab represents the two-digit number $(10a + b)$, and x and y are the weights (cwt) of the small and large bags, respectively. The only solutions occur for

$$y = 32x$$
$$8x + 5y = 4 \tag{6.2}$$
$$6x + 9y = 7$$

Here, the triplet (c, d, e) and (f, g, h) are interchangeable.

This is a previously published puzzle of mine [1] and serves as a good example of a problem in which a solution is possible only by assuming a particular set of conditions.

Reference

[1] Clarke, Barry R. *Brain Busters: Mind Stretching Puzzles in Math and Logic*. New York: Dover Publications Inc., 2003, p.76.

7 Logic

7.1 The four chests

King Midas has sacked his treasurer Manny Baggs but decides to give him the chance of a golden handshake. So he presents him with four treasure chests, each of which has a statement attached to the side which refers to the location of a thousand gold coins. Only one chest contains the coins and if Mr Baggs can identify the correct chest then he can leave his job a rich man. The statements are as follows.

Statement A: "Neither A nor D."
Statement B: "A chest at the end of the line."
Statement C: "A chest adjacent to this."
Statement D: Either A or B."

Exactly one statement is true.

Which chest holds the gold coins?

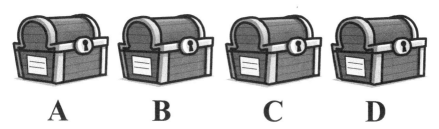

Figure 7.1   The choices facing Manny Baggs.

DOI: 10.1201/9781003358275-7

7.2 Carb heavy

Alerted by the burger alarm, Mildred Flannel has just witnessed four robbers run out of her local fast-food shop with armfuls of french fries. At Copham and Shopham police station, she is now making a drawing of the faces A to D. Although the items in each row of hair, eyes, and mouth are correctly drawn, only one item in each row is correctly positioned. The following facts are true about the correct order.

(1) Neither hair A nor hair C is at C.
(2) Eyes A are not at B.
(3) Mouth C is one place to the left of eyes D.
(4) Neither B nor C's mouth is at A.
(5) Eyes A are one place to the right of hair A.

Can you reconstruct the faces correctly?

Figure 7.2   The identity parade.

7.3 Movie mutations

There are nine make-up and costume cubicles on a movie set. The cubicles are labelled A to I and are surrounded by movie extras. Each extra on the left has passed through the three cubicles directly to her right and has been transformed into the person on the far right (e.g. the person to the left of A has passed through cubicles A, B, and C to finish looking like the person to the right of C). Similarly, each extra shown above the cubicles has passed through the three directly below to finish looking like the person at the bottom. Each cubicle always produces a single change only to the person passing through (and each cubicle must cause the same change to both people that enter); compelling her to add or remove her hat, add or remove her necklace, change her hair colour to white or grey, produce a sad or happy mouth, or induce her to open or close her eyes. No change is repeated in the same row or column.

What does a visit to each cubicle do?

Figure 7.3   The face alterations as they pass through the cubicles.

7.4 Gorilla warfare

It is Christmas Day and amongst the coastal trees of equatorial Africa the apes are singing Jungle Bells. Tarzan is still only a lad, and as a Christmas present, they decide to show him the ropes. So a family of four gorillas and a family of four orangutans turn up ready to reveal their tree-swinging technique. However, the two families cannot agree on the finer points, and so the whole affair descends into a banana fight. The skirmish alternates between a gorilla and an orangutan throwing a banana at a member of the other family. A gorilla hits an orangutan with a banana, who then strikes a gorilla with a banana, and it continues alternately, until all eight apes have been hit. Altogether eight bananas are thrown, each ape throws only one banana, and each ape is struck by only one banana. The following facts apply.

(1) Ron hits the orangutan who hits the gorilla who hits Celia.
(2) Stan hits the orangutan who hits the gorilla who hits Benny.
(3) Celia hits the gorilla who hits the orangutan who hits Toby.
(4) Arthur hits the gorilla who hits the orangutan who hits Peggy.

Who hits Denise?

7.5 Crazy cards

Seymour Foulke is attending the Lonely Logicians Club whose logical slogan is "either you are alone or U.F.O." Unfortunately no one else has turned up. Anticipating a less than full attendance, the group leader Mike O' Nundrum has left a problem on the noticeboard should anyone summon up the will to attend. Three cards are set out in a line as shown. Each has a statement written on it indicating the correct position of the card in the line. At most, one of the three statements is true.

Can you find the correct order?

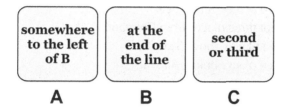

Figure 7.4   The positional statements on the three cards.

7.6 Cold reception

After holding a house-warming party at his new igloo, Frosty the snowman is now homeless. So his friends invite him to stay at their accommodation. Frosty has listed his top five preferences, but since his brain is made of ice he has written them down in the incorrect order. The first names, surnames, and locations of his potential hosts are shown above. Although each item is in the correct column, only one item in each column is in the correct position. The following facts are true about the correct order.

(1) Cat is two places above Colorado.
(2) Neither Seal nor Yakutsk is fourth.
(3) Funky is two places below Bear.
(4) Alaska is not three places from Colorado.
(5) Holly is two places above Seal.
(6) Sally is three places below Alaska.

Can you find the correct first name, surname, and location for each accommodation?

Table 7.1 A five rows and three columns mix-and-match

	first name	surname	location
1	Adah	Bear	Alaska
2	Bob	Cat	Colorado
3	Holly	Hare	Greenland
4	Funky	Fox	Snag
5	Sally	Seal	Yakutsk

7.7 Parking problems

Noah C. Parker, the car parking attendant is driving his customers up the wall with his demands. The car park is divided into a 4 × 4 grid of parking spaces, the same 16 cars park there every day, and Noah is very strict about which space each car should occupy. His rules are as follows.

(1) The Chrysler is two below and one to the right of the Kia.
(2) The Bentley is one place horizontally to the left of the Maserati.
(3) The Infiniti is one place horizontally to the left of the GMC.
(4) The Subaru is one to the right of and one above the Lexus.
(5) The Kia is one place vertically above the Bentley.
(6) The Audi is one place horizontally to the left of the Jeep.
(7) The Honda is one place vertically below the Genesis.
(8) The Porsche is two horizontally to the right of the Ferrari.
(9) The Jeep is one vertically above the Lexus.
(10) The Toyota is one to the right of and one below the Genesis.

Where is the Nissan parked?

7.8 Monkey business

Stoney Livingstan was amazed to chance upon a group of English-speaking monkeys in the amazon jungle. Dinner was about to be served and Uru their food server was holding a can-opener.

"You don't need a can-opener to peel a banana," said Stoney.

"I know," said Uru, "this is for the custard."

Meanwhile, five monkeys in the dinner queue were trying to ascertain their agreed order. Capa and Ethel, who never stood next to each other, were arguing about it.

(a) "I think it's Adu, Booba, you, me, and Dodo," said Capa. However, only two of them were correctly positioned.
(b) "Well, I think it's Adu, Dodo, you, Booba and me," said Ethel. This time only one was correctly placed.
(c) Failing to agree, they eventually stood in the order Adu, Dodo, Ethel, Booba, and Capa. This time exactly two were correctly placed.

What was their agreed order?

7.9 Patternbots

Five robots from the Mindless Metal Factory are shown on display A to E. Each robot has three variable characteristics (head, body, and legs) and five different patterns for each characteristic (white, grey, black, vertical stripes, horizontal stripes). However, although each pattern is in the correct row, due to a fault in the design software, only one of the five patterns in each row is correctly positioned. The following facts are true about the correct order of patterns.

(1) Legs B are somewhere to the left of head D which is not to the left of legs C.

(2) Body A is one place to the right of head C and one place to the left of legs A.

(3) Body C is not at D.

(4) Head A is two places to the left of body E and one to the left of legs D.

Can you reconstruct the designs correctly?

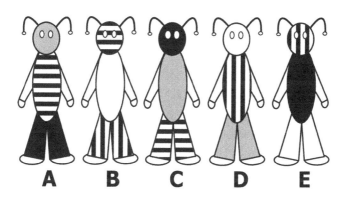

Figure 7.5   The incorrect patterns on the five robots.

7.10 The three hats

Agatha, Beryl, and Celia have decided to give up their materialistic life-style, sell their houses, and live a life of liberty on the streets. Since their savings are to be left untouched in the bank, each decides to buy a hat to collect money from passers-by. In the milliner's shop, the shop assistant Kanye Elpmee has the mistaken impression that they want a hat to *wear*, so he places one on each person's head without the recipient knowing what colour it is. The three friends know that there are only six hats that are suitable for begging on the street: one is black, two are grey, and the other three are white. Each person can only see the other two hats being worn and cannot see which hats remain unused. From what she can see, Agatha reasons "My hat is one of two colours," Beryl deduces that "my hat is one of three colours," and Celia works out that "my hat is one of three colours."

What colour hat is each person wearing?

7.11 Unlocking the mind

Harry Cowardini the escapologist has visited his doctor complaining of claustrophobia. So his doctor has told him he needs to get out more. As a result, Harry has decided to abandon his locked-trunk routine and change to a mind-reading act. His newly appointed assistant is a master of logic. In one routine, Harry is blindfolded and a member of the audience is invited to hold up a handkerchief, a key fob, a wallet, a pair of spectacles, or an iPhone. This is then to be concealed by a hat, a jacket pocket, a trouser pocket, a bag, or a scarf. His assistant then makes three seemingly random double-statements, the first of each pair being about the chosen article and the second about the method of its concealment. During one performance, the three double-statements are as follows:

(1) "A handkerchief or wallet" and "A jacket pocket or bag."
(2) "A key fob, spectacles or iPhone" and "A hat, jacket pocket or trouser pocket."
(3) "A key fob or iPhone" and "A jacket pocket or bag."

Now, Harry and his assistant have agreed in advance that in each double-statement there is to be one true and one false assertion, although the order will be unknown.

What item is chosen and where is it concealed?

7.12 Musical chairs

At a birthday party, six chairs numbered 1 to 6 are arranged sequentially in a circle for a game of musical chairs. When the music stops, six bottoms park themselves on six chairs, each chair being occupied by one person only. When seated, the players face inwards and the person whose birthday it is manages to sit in chair 1. The positions in the circle are as follows.

(1) Malcolm, who does not have the birthday, sits immediately to the right of Sally, who is not opposite the birthday person.
(2) Jennifer does not sit next to Uri.
(3) Nat is the first to sit down.
(4) Victor sits two places to the right of Jennifer.
(5) Uri sits at least two places from the birthday person.

Whose birthday is it?

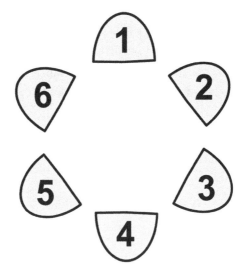

Figure 7.6   The circle of chairs.

7.13 Cards and colours

At the School of Logic, the votes have been counted in the election of their new president. The outgoing president, Professor Neuron, has placed three cards on the table that reveal the first three places (first, second, third). Each card has a candidate's name written on it face down (Brian Box, Finn King, Alby Quick). On each uppermost face there are two statements: one about that candidate's final ranking in the election, and one about the face down name.

The black card states, "This card has a final ranking somewhere above a card with a true positional statement" and "This card is neither Brian Box nor Finn King."

The grey card states, "This card is not adjacent in the rankings to a card with a false positional statement" and "One of the other cards has the name Alby Quick."

The white card states, "This card is neither second nor third" and "Finn King is not second."

One of the cards has two true statements, one has two false, and the other has one of each.

What colour card is first and what is the name on it?

7.14 The widow's woe

When the old miser Titus Zell died he left no money to his wife, and had signed over his house to Moping Moggies a rescue centre for cats. Even Scratter the family dog was left with no budget for food. So his widow had to contact her five sons to see if they could help out. Unfortunately, they had inherited their father's thrift and anyone who had money lied not only about possessing wealth but also about what the other sons possessed and had said. Liars contradicted the statements of others and those who had no money always told the truth.

(a) Asspot reported to his mother that Catwinge had said, "Exactly one of my four brothers has money."
(b) Bogwash claimed that Eyesore had said, "Exactly two of my brothers have money."
(c) Catwinge told his mother that Duckpan had said, "Exactly three of my four brothers have money."
(d) Duckpan reported that Bogwash had told him, "All of my four brothers have money."
(e) Eyesore said that "Asspot claims to have money" and "Catwinge has money."

Which sons had money?

7.15 Scholar mutations

Twelve sets of academic books are labelled A to L. They are surrounded by scholars who are somehow changed by reading them. Each person on the left has read the three piles directly to his right and has been transformed into the person on the far right (e.g. the figure to the left of A has read sets A, B, and C to finish as the person to the right of C). Similarly, each scholar shown above the books has read all the piles of books directly below to be transformed into the person at the bottom. Each set always produces a single change only to the person reading (and each set must cause the same change to both people that read them); compelling him to add or remove his mortar board, add or remove his glasses, change his hair colour to white or grey, produce a sad or happy mouth, or induce his moustache to fall off or grow back. No change is repeated in the same row or column.

What does each set of books do?

Figure 7.7   The face alterations as they read through the piles of books.

7.16 Taxi rank

At the Kickem-n-Lickem kung fu club, six practitioners are waiting for four taxis to arrive. Their belt ranks in descending order are black, brown, purple, blue, green, and orange. When the taxis arrive, only four of the six take a car, one to a taxi. Numbering the taxis 1 to 4 corresponding to their order of arrival, the following facts are true about who climbs into one.

(1) Taxi 2 is taken by a higher belt rank than the person in taxi 3.
(2) The person who takes taxi 1 is two positions away in the given order from the person who takes taxi 3.
(3) The blue belt and the green belt are not in consecutive taxis.
(4) Taxi 4 is taken by a person three positions away in the given ranking from the one who took taxi 2.
(5) The person who takes taxi 1 has a lower rank than the one in taxi 4.

Can you find the belt colour for the person in each taxi?

7.17 Lying low

Tired of being asked for selfies whenever they go out, the Seven Dwarfs have each changed their personal name, house name, tunic colour, and hat colour. However, when they march off to work in line, they still do so in a specific order. The dwarf now known as Denser has written this down but he has made many errors. Although each item is in the correct column, only one entry in each column is correctly positioned. The following facts are true about the correct order.

(1) Dungroovin is one below the red hat and two above the silver tunic.
(2) The crimson tunic is somewhere above the yellow hat.
(3) The violet hat is two below the lilac tunic and two above Achoo.
(4) Denser is not above Binfinkin.
(5) Outasugar is one above Snoozey who is two above Prof.
(6) The gold tunic is three above the green hat which is two below Luvloungin and three below Misery.
(7) The indigo hat is one below Funfoolin and one above the turquoise tunic.

Can you find the correct dwarf name, house name, tunic colour, and hat colour for each position?

Table 7.2 A seven rows and four columns mix-and-match

	name	house	tunic	hat
1	Achoo	Binfinkin	amber	red
2	Denser	Dungroovin	beige	orange
3	Jolly	Funfoolin	crimson	yellow
4	Misery	Gonfishin	gold	green
5	Prof	Luvloungin	lilac	blue
6	Snoozey	Outasugar	silver	indigo
7	Thumper	Whywurrie	turquoise	violet

7.18 Manhattan

Four friends, Antwit, Babble, Cringe, and Dibdib, decide to visit the Man-Hat-On shop in Manhattan, New York. There are six hats to choose from—three white and three black—and the white hats have the magic property of turning any wearer into a habitual truth-teller while a black hat has no such restriction, allowing the wearer to lie or tell the truth. Each person is blindfolded and a hat is placed on his head. They are then positioned at the corners of a square, the blindfolds are removed, and with the remaining hats out of sight, each can see the other three hats but not his own. Each is now asked to make a statement about the colours of the three hats he can see.

Antwit says, "two black and one white."
Babble says, "one black and two white."
Crumble says, "at least two black."
Dibdib says, "one black and two white."

At least one of them is lying.

Can you give the colour hat each is wearing and name any that are lying?

Solutions

7.1 The four chests

The coins are in chest C and statement A is true. If A is true the coins are at B or C. If B is true then A or D. If C is true then B or D. If D is true then A or B. With one statement true we need a letter that appears once only in this list which is C.

7.2 Carb heavy

Figure 7.8   The correct drawings of the identity parade.

7.3 Movie mutations

A hair grey	**B** hair white	**C** add hat
D remove necklace	**E** happy mouth	**F** remove hat
G open eyes	**H** add necklace	**I** close eyes

7.4 Gorilla warfare

Toby hits Denise and the circular order is RASCPBTDR. The orangutans are Arthur, Benny, Celia, and Denise. The gorillas are Ron, Stan, Toby, and Peggy. From conditions (1), (3), and (4), using first letters of names, there are two possibilities:

AR_orangutan_PC_gorilla_orangutan_T
and
R_A_gorilla_CP_orangutan_T

Condition (2) only fits in with the latter (for the former, one ape must be hit twice which is invalid). So we have RASCPBT. This means that Denise can only fit after T (and before Ron because we have a circular chain) so Toby hits Denise (who must then hit Ron).

7.5 Crazy cards

The card order is C, B, A with no true statements. "At most one" statement means zero or one. Let us suppose that there is exactly one true statement. Let us suppose that it is A. Then it can be at A1 or A2, where the number indicates its position. Cards C and B are false so C can be at C1 and B can be at B2. However, no card can be third. Suppose instead that it is C that is true. Then we have C2 or C3. Cards A and B must be false and so B can be at B2 while A can be at A2 or A3. However, no card can be first. For the final case, suppose B to be true. Then B is at B1 or B3. If B1, then the false C is also at C1 (invalid). If B3, then the false A cannot be to the right of B (invalid). So there cannot be exactly one true statement. If we assume no statements to be true ("at most one"), then we can have C1B2A3 without contradiction.

7.6 Cold reception

Table 7.3 Solution to the five rows and three columns mix-and-match

	first name	surname	location
1	Bob	Cat	Alaska
2	Adah	Fox	Yakutsk
3	Holly	Bear	Colorado
4	Sally	Hare	Greenland
5	Funky	Seal	Snag

7.7 Parking problems

The Nissan is in the bottom left-hand corner of the car park.

Table 7.4 Solution to the 4×4 grid parking problem

Genesis	Ferrari	Kia	Porsche
Honda	Toyota	Bentley	Maserati
Audi	Jeep	Subaru	Chrysler
Nissan	Lexus	Infiniti	GMC

7.8 Monkey business

The correct order was Capa, Adu, Ethel, Booba, and Dodo. Using first letters of names, let A1 denote Adu in position 1 from front to back. From the change from (b) to (c), either E3 or C5 is correct in (c). We now check the four cases of A1 correct then incorrect with E3 correct or C5 correct.

(i)   Assuming A1 is correct, then D2, C3, B4, E5 in (b) are all wrong. Assuming E3 is right, then C5 is wrong in (c), and B2, C4, D5 are wrong in (a). This gives A1, C2, E3, D4, B5 violating the condition that C and E are not together.

(ii) Assuming A1 is correct and E3 is wrong in (c) instead of C5, so C5 is correct, then D2, E3, B4 are wrong in (c). It follows that with E3, C4, D5 wrongly positioned in (a) then B2 must be right. This gives A1, B2, D3, E4, C5, and C and E are again wrongly together. So our assumption that A1 is correct must be wrong.

(iii) Let us assume that A1 is wrong, C5 is correct with E3 wrong. Then C4 and D5 are wrong in (a) and B2 is correct. This means there is only one correct place in (a) instead of two.

(iv) Let us assume that A1 is wrong, E3 is correct and C5 is wrong. Then C3 and E5 are wrong in (b) and either D2 or B4 is correct. Assuming D2 is correct, then B2, D5 are wrong in (a) so we have E3, C4, violating the CE separation condition. Assuming B4 is correct instead of D2, then B2 and C4 are wrong in (a) so D5 must be right, and A2 is the only position for A. We can now only have C1, A2, E3, B4, D5.

7.9 Patternbots

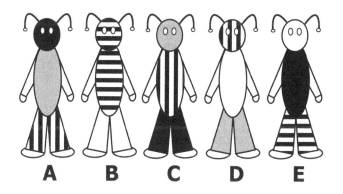

Figure 7.9   The correct pattern arrangements for the robots.

7.10 The three hats

Agatha has white, Beryl has grey, and Celia has grey. The claim "one of two colours" can only arise if one of the other two is black (leaving grey or white for herself) or the other two are grey (leaving black or white for herself). The claim "one of three colours" means grey and white, or two white. With two people making this claim there can be no black in the three hats. So Agatha has seen two grey, and both Beryl and Celia have seen a grey and a white, with Agatha wearing a white hat.

7.11 Unlocking the mind

The spectacles were placed in a bag. Since "jacket pocket or bag" appears in (1) and (3), these two statements are either both true or both false. If they are both false, then "A handkerchief or wallet" and "A key fob or iPhone" are both true which is not possible as they are inconsistent (no shared item). So they must be both true and from (1) and (2), their false first statements point to the spectacles. This means that the first statement in (2) is true and "A hat, jacket pocket or trouser pocket" is false ruling out the jacket pocket. So from the true second statements in (1) and (3) the bag is used for concealment.

7.12 Musical chairs

Jennifer is the birthday person in chair 1, Malcolm is in 2, Sally is in 3, Uri is in 4, Victor is in 5, and Nat is in 6. From (1), Malcolm sits immediately to Sally's right, and from (4), Victor sits two places to the right of Jennifer. Viewing clockwise, this allows MSV_ J _ or MS_V_J. Considering (2), this allows only MSUVNJ. Using (5) to identify the birthday person, it is J, M or N. Condition (1) rules out M who does not have a birthday, and also eliminates N who is opposite Sally. So Jennifer has the birthday in chair 1.

7.13 Cards and colours

First is Alby Quick, second is Brian Box, and third is Finn King. The white card is first but the order of the grey and black is undetermined (which matters not as far as this problem is concerned).

Let us consider the four true (T) and false (F) permutations for the positional statements of the white and grey cards (TT, FT, FF, TF, respectively).

(i)   If we have white T and grey T, then black must be F (because we cannot have three T positional statements). Then white T is first and grey T must be next to black's false positional statement (invalid).

(ii) If we have white F and grey T, we must have grey T first and white F third (since grey cannot be adjacent to the false white) leaving black second. Grey's T statement means that black can only be T but then it is not above a card with a T statement (invalid).

(iii) If we have white F and grey F, then black must be T (for two T to appear on one card). However, black is not then positioned above a card with a T positional statement because the other two are F (invalid).

(iv) This only leaves white T and grey F, so white T is first and if grey F is to be satisfied then we must have black F. In order of position (first, second, third), this gives two solutions: (1) white T, grey F, black F; and (2) white T, black F, grey F. The truth status of the three colours is consistent whatever their positions and we know white is first.

If we now turn to the name statements on the cards: a name statement T must be paired with white's placement statement T (otherwise there is no card with two T), and so the only possibilities in the second statements for black, grey, white are (a) T, F, T or (b) F, T, T.

For (a) the consequences are as follows.

Black name statement T: the black card is A
Grey name statement F: the grey card is A
White name statement T: F is first or third

Clearly, two cards cannot be A. For (b) the consequences are as follows:

Black name statement F: the black card is B or F
Grey name statement T: the grey card is B or F
White name statement T: F is first or third

White is not F because the name F must be one of grey or black. So white which is first has name A. This means F is not first and must be third. This leaves B second.

We can summarise the results for the three card colours in Table 7.5.

Table 7.5 Summary of conclusions

	black	grey	white
Positional statement	F	F	T
Name statement	F	T	T
Position in line	2 or 3	3 or 2	1

7.14 The widow's woe

The brothers who lie and have money are Asspot, Bogwash, and Eyesore. Having money means lying so in Eyesore's statements, Asspot could not claim to have money whether he had some or not. A truth-teller would have none to declare and a liar would conceal having some and claim none. So Eyesore is a liar (E_L) and Catwinge has no money and is a truth-teller (C_T).

With C_T and E_L, let us examine the case that Duckpan lies (D_L). Then Catwinge accurately reports Duckpan's lie in (c), so none, one, two, or four of A, B, C, E lie. From what is already known and assumed, the possibilities are $A_T B_T C_T D_L E_L$, $A_L B_T C_T D_L E_L$, and $A_T B_L C_T D_L E_L$. From (a), knowing C_T, if A_T then we have two of the above three possibilities $A_T B_T C_T D_L E_L$ and $A_T B_L C_T D_L E_L$. However, we can see that it is not true that only one of A, B, D, E lie (have money). So we must have A_L and the remaining possibility $A_L B_T C_T D_L E_L$. From (b), with $A_L B_T C_T D_L E_L$, and B_T and E_L it is not true that exactly two of ABCD are liars. So there is an inconsistency assuming D_L.

With C_T and E_L, let us instead assume D_T. From (c), exactly three of ABCE lie. So we must have $A_L B_L C_T D_T E_L$. We must now check for consistency with the other statements. With A_L and C_T, (a) is false which works. With B_L and E_L, (b) can be the false report of a false statement which works. With B_L and D_T, (d) is false which works. So we conclude that $A_L B_L C_T D_T E_L$.

7.15 Scholar mutations

There must be pairs of changes that are invisible because one reverses the other. The members of a pair will appear together in the same row or column. We must have the same set of changes for the rows and columns. So the hidden pair in the third row must be hair white and hair grey, while the hidden pairs in the first and second columns are add and remove board, and add and remove moustache. In the third column, F must be remove glasses

Table 7.6 The effect that a book pile has on a scholar reading it

A add glasses	B remove board	C hair grey
D remove moustache	E sad mouth	F remove glasses
G hair grey	H hair white	I add moustache
J add moustache	K add board	L happy mouth

and L must be happy mouth, there being only one row containing these. In the second row, taking account of the known hidden pairs in the first two columns, sad mouth can only be at E leaving remove moustache at D. So add moustache must also be in the first column leaving add and remove board in the second. In the third row, add board is only in the second column at K. So J is add moustache. Then add moustache in the third column is in the third row at I, leaving C as hair grey. In the first row, add glasses is only in the first column at A, so B is remove board, H is hair white, and G is hair grey.

7.16 Taxi rank

The four taxis, in their order of arrival, were taken by the green, black, purple, and blue belts. From (1) and (5), the black belt didn't take taxi 1 or 3, and the orange belt didn't take taxi 2 or 4. From (4), the ones in taxis 2 and 4, in some order, must have been either the black belt and blue belt, or the brown belt and green belt. This eliminates purple from these taxis. From (3), if the blue belt is in taxi 2 or 4 then neither blue nor green can be in taxi 3, and if the green belt is in taxi 2 or 4 then neither green nor blue can be in taxi 2 or 4. So neither blue nor green is in taxi 3. From (2), we can eliminate brown, purple, and orange from taxi 1, leaving only the blue belt or green belt. Using (3), if blue is in taxi 1, both blue and green are eliminated from taxi 2, and if green is in taxi 1, the same applies. So taxi 2 can have either black or brown. We now know that either black is in taxi 2 with blue in taxi 4, or brown is in taxi 2 with green in taxi 4. So black and brown can be eliminated from taxi 4. The table of possibilities that remain is now shown.

From (5), the green belt must be in taxi 1 and the blue belt in taxi 4, so using (2), the black belt took taxi 2, and using (2) and the purple belt took taxi 3.

Table 7.7 Taxi rank deductions

taxi 1	taxi 2	taxi 3	taxi 4
blue	black	brown	blue
green	brown	purple	green

7.17 Lying low

Table 7.8 Solution to the seven rows and four columns mix-and-match

	name	house	tunic	hat
1	Jolly	Outasugar	amber	orange
2	Snoozey	Funfoolin	lilac	red
3	Misery	Dungroovin	gold	indigo
4	Prof	Luvloungin	turquoise	violet
5	Denser	Binfinkin	silver	blue
6	Achoo	Gonfishin	crimson	green
7	Thumper	Whywurrie	beige	yellow

7.18 Manhattan

Antwit wears black, Babble has black, Crumble has white, and Dibdib wears black. There are two liars: Babble and Dibdib.

Anyone wearing a white hat must be a truth-teller, but a black hat wearer can be either a truth-teller or a liar. So a liar can only have a black hat. A truth-teller can have either. Let us also use the first letters of names.

Suppose Antwit is a liar, wears a black hat, and does not see BBW (any order). With the order ABCD, the reality can then be BBBB, BBWW, BWBW, BWWB, or BWWW. The first is ruled out because there are only three black hats. The second is ruled out because, with a white hat, Dibdib is a truth-teller and should see one black and two white. The third is wrong because Babble, with a white hat, tells the truth and should see one black and two white. We can eliminate the fourth because again, with a white hat, Babble is a truth-teller and should see one black and two white. Finally, the fifth cannot be correct because Crumble, with a white hat, tells the truth and sees at least two black hats. This means that Antwit must tell the truth and sees _ BBW, _BWB, or _WBB.

Suppose Babble tells the truth, and sees W_BW, W_WB, or B_WW. Combined with Antwit, we can have either WBBW or WBWB. In the first, Dibdib wears a white hat, is a truth-teller, and sees one black and two white which is wrong.

In the second, Crumble wears white, tells the truth, and sees at least two black so WBWB is feasible. Dibdib cannot tell the truth because we already have three truth-tellers (Antwit, Babble, and from WBWB also Crumble with a white hat) and there must be at least one liar. However, if Dibdib tells lies then he fails to see a BWW combination so that doesn't work. The consequence is that contrary to our initial assumption, Babble must be a liar and wears a black hat. From Ant's possibilities, this allows _BBW or _BWB. However, Babble now sees either three white, three black, or two black and one white and so we have either BBBW or BBWB.

In the first case BBBW, Dibdib has a white hat and must be a truth-teller. This means Dibdib sees two white and one black which does not work. In the second case BBWB, Crumble is a truth-teller and sees at least two black which works. If Dibdib tells the truth, he sees one black and two white which is wrong. So he must be a liar and sees either three white, three black, or two black and one white. So BBWB works and this gives the complete solution.

Visual-lateral

8.1 The dead dog

Shown is a smiling dog constructed from matchsticks. It is lying down, facing to the left, with its legs stretched out and its tail in the air. One matchstick must be moved to another part of the figure so that the dog looks dead.

Which matchstick must be moved?

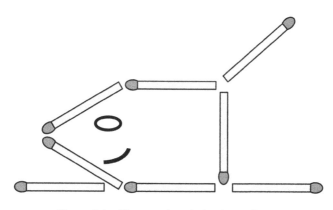

Figure 8.1   The matchstick dog lying down.

DOI: 10.1201/9781003358275-8

8.2 Spot the dice

Shown is a line of four dice in their box. The spots on the upper face of the fourth dice are missing.

How many spots have been erased?

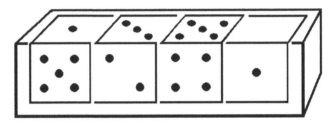

Figure 8.2   The four dice in their box.

8.3 The tin door

While exploring a cave, two adventurers unexpectedly found a locked tin door set into a wall. On the adjacent wall was a set of curious fractions. Deciphering the message reveals what is in the room behind the door.

What is in the secret room?

Figure 8.3   The cryptic message on the wall.

8.4 The broken timepiece

In the diagram above, there should be only one clock that shows the correct time. To find it, the seven circles can be rearranged among the seven positions shown. However, they must not be rotated.

What is the correct time?

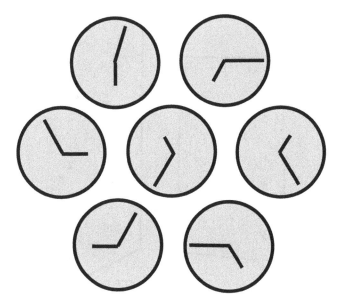

Figure 8.4   Seven circles that look like clocks.

8.5 Rough graph

The graph above is incomplete.

What large capital letter should be placed to the right of the lower left point?

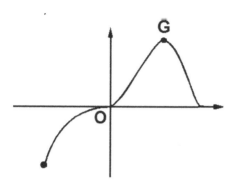

Figure 8.5   The graph with a laugh.

8.6 Duck to cat

As Whiskers the cat was enjoying a walk one Saturday afternoon, he spied Dotty the duck in the nearby pond. Under cover of the wall at the top end of the pond, Whiskers tiptoed towards the water. He waited for the right moment, then suddenly popped up from behind the wall and scared Dotty away. Dotty is shown above with her mouth open in fright.

Can you rearrange two matches to change the duck into a cat?

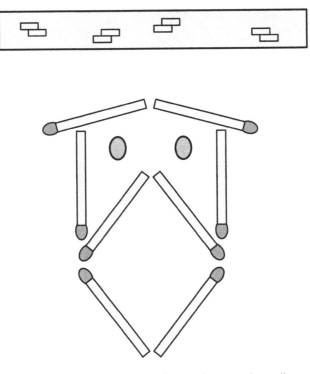

Figure 8.6   Duck sitting in the pond next to the wall.

8.7 The impossible anagram

Shown are seven letters which you must *not* try to rearrange! If you make any attempt to do so then you've cheated. So, without rearranging the letters, there is still a way to solve the anagram.

How is it possible?

Figure 8.7   The letters of the anagram.

8.8 The hungry fish

A shoal of fish were enjoying a swim when the fish at X decided to eat the five fish A–E. This it achieved by moving only along the straight lines, visiting each position once only and finally returning to X to swim along with the shoal. The fish at C was eaten some time before D who was not the last eaten. Fish B was devoured some time before A. The route that was taken did not cross over itself at O.

Can you draw the route that the hungry fish took?

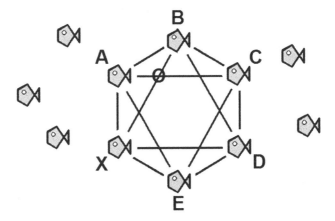

Figure 8.8   The shoal of fish.

8.9 The four doors

Shown are four glass doors.

(1) Door A has a letterbox and a push-bar.
(2) Door B has an oval window.
(3) Door C has a letterbox and a cat flap.
(4) Door D has a rectangular window and a push-bar.

If one of the doors is fully opened a car will be revealed.

Which door is to be opened?

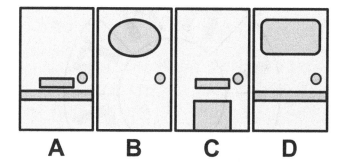

Figure 8.9   The four glass doors.

8.10 Hail clip!

Alf and Bert were sitting in the pub pondering over a ring-shaped cardboard beer mat.

Said Bert, "how can you move half of the letters to form the words HAIL CLIP reading clockwise?"

With that, Alf produced a pair of scissors and with eight radial cuts and a subsequent rearrangement of the letters completed the task.

"OK," said Bert, "but I can do it with only one continuous cut."

How is it possible?

Figure 8.10   The HAIL CLIP beer mat.

Hints

8.1 *The dead dog*
Try looking at it upside down.

8.2 *Spot the dice*
If you look closely, the dice appear to be something else.

8.3 *The tin door*
How could the shiny tin be useful in solving this?

8.4 *The broken timepiece*
There should be only one clock, rather like a jig-saw puzzle.

8.5 *Rough graph*
Try turning your head counter clockwise.

8.6 *Duck to cat*
We don't see the cat behind the wall but the water helps us see it.

8.7 *The impossible anagram*
The temptation to cheat is enormous! Do you have a mirror?

8.8 *The hungry fish*
There is only one solution to the logic puzzle because the fish ends up swimming with the shoal.

8.9 *The four doors*
If a door can be fully opened it can be superimposed on an adjacent one.

8.10 *Hail clip!*
An octagonal cut is needed.

Solutions

8.1 The dead dog

Move the front leg to the top of the figure as shown. The dog is now lying on its back with its legs in the air. Viewed upside down, the closed mouth is now a closed eye and the open eye is now an open mouth. The dog looks dead.

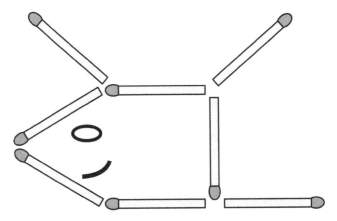

Figure 8.11   The matchstick in its new position.

8.2 Spot the dice

Two spots are missing. If one inspects the four dice closely it can be seen that they are constructed from the digits 6, 5, 9, and 3. Each digit gives the total of spots on that dice. Since there is already one spot visible there must be two missing.

8.3 The tin door

Reflecting the message in the tin door reveals "tin mine inside it." See Figure 8.12.

tIN mIN eIN SID EIt

Figure 8.12   The result of reflecting the fractions in the tin door.

8.4 The broken timepiece

As the puzzle title suggests, the clock has been broken into pieces. When the pieces are rearranged as above, they form a hexagonal clock showing the time 6:05.

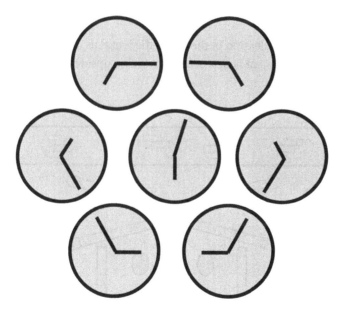

Figure 8.13   The reconstructed hexagonal clock.

8.5 Rough graph

A large letter D gives the dog's ear. The dog's head is looking upwards.

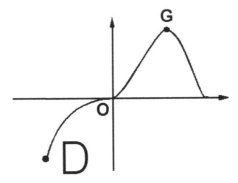

Figure 8.14   The picture of a dog.

8.6 Duck to cat

The view is of the reflection of the cat in the water as its head pops up from behind the wall. The cat is best seen by viewing upside down.

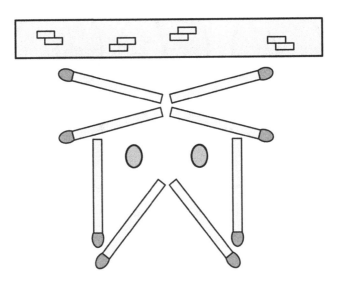

Figure 8.15   The reflection of the cat in the water.

8.7 The impossible anagram

Place a mirror above the letters (the black screen of an iPhone should suffice) to see the solution. All the letters of "reflect" are present in the original set of letters and reflecting them produces this word. Paradoxically, one needs to know the answer before one has a method of finding the answer!

8.8 The hungry fish

The correct route traces out the shape of the fattened fish swimming with the shoal. There are two possible solutions: the one shown XECDBAX and the left-to-right reflection of it XBCDEAX. However, the puzzle states that the fish ends up swimming with the shoal, so this eliminates the latter reflected solution.

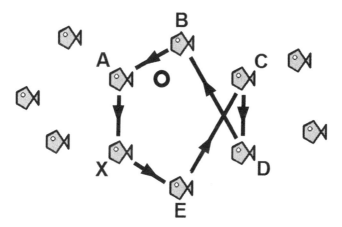

Figure 8.16   The enlarged fish.

8.9 The four doors

Door D is to be fully opened. Since it is made of glass, when it is swung wide open its features are superimposed on those of door C which can be seen through it.

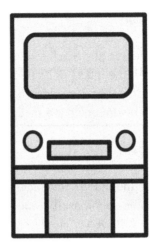

Figure 8.17   Door D superimposed on door C.

8.10 Hail clip!

Make an octagonal cut as shown and rotate the outer ring one letter counter clockwise. How does this constitute a movement of half of the letters? The rotation moves the top half of the letters!

Figure 8.18   The octagonal cut.

Appendix

Computer program for 6.4. The five digits
'Written in Liberty Basic (developed by Carl Gundel, 1991).

```
for p1=1 to 9
for p2=1 to 9
for p3=1 to 9
for p4=1 to 9
for p5=1 to 9
' No two digits the same
if p1 < > p2 and p1 < > p3 and p1 < > p4 and p1 < > p5 and
p2 < > p3 and p2 < > p4 and p2 < > p5 and p3 < > p4 and p3
< > p5 and p4 < > p5 then
 ' Exactly two digits are square
   X=0
   ' First digit
   if p1 = 1 or p1 = 4 or p1 = 9 then
    X=X+1
   end if
   ' Second digit
   if p2 = 1 or p2 = 4 or p2 = 9 then
    X=X+1
   end if
   ' Third digit
   if p3 = 1 or p3 = 4 or p3 = 9 then
    X=X+1
   end if
```

```
' Fourth digit
if p4 = 1 or p4 = 4 or p4 = 9 then
 X=X+1
end if
' Fifth digit
if p5 = 1 or p5 = 4 or p5 = 9 then
 X=X+1
end if
 if X=2 then
 else
    goto 30
 end if
' First and fifth differ by 2
 if abs(p1 - p5) = 2 then
 else
   goto 30
 end if
 ' Second and third sum to fourth
   if p2 + p3 = p4 then
   else
     goto 30
   end if
' First and second differ by 1
if abs(p1 - p2) = 1 then
   else
     goto 30
   end if
  ' The second and fourth differ by one
   if abs(p2 - p4) = 1 then
   else
     goto 30
   end if
' Fifth is odd
   if p5 = 1 or p5 = 3 or p5 = 5 or p5 = 7 or p5 = 9 then
   else
     goto 30
   end if
```

```
else
 goto 30
end if
20 print p1; p2; p3; p4; p5
30
next p5
noxt p4
next p3
next p2
next p1
end
```

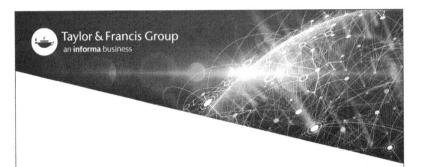

Printed in the United States
by Baker & Taylor Publisher Services